A Book Of

SOLID STATE PHYSICS

T.Y.B.Sc. Physics : PH – 332 : Semester-III
As Per New Revised Syllabus
With Effect from June 2015

Dr. S. D. AGHAV
M.Sc., M.Phil., Ph.D.
Ex. Vice Principal,
Baburaoji Gholap College,
Sangavi, **PUNE**.

B. M. LAWARE
M.Sc., M. Phil.
Head, Department of Physics,
Prof. Ramkrishna More,
Arts, Commerce & Science College,
Akurdi Pradhikaran, **PUNE**

Dr. P. S. TAMBADE
M.Sc., Ph.D.
Head of Physics,
Prof. Ramkrishna More,
Arts, Commerce & Science College,
Akurdi, **PUNE**.

V. K. DHAS
M. Sc., M. Phil.
Ex. Head, Department of Physics,
New Arts, Commerce and Science College,
AHMEDNAGAR

N1852

T.Y.B.Sc. : SOLID STATE PHYSICS (S-III)

ISBN 978-93-51645-87-0

Fifth Edition : **August 2019**

© : **Authors**

Published By :
NIRALI PRAKASHAN

Abhyudaya Pragati, 1312, Shivaji Nagar,
Off J.M. Road, PUNE – 411005
Tel - (020) 25512336/37/39, Fax - (020) 25511379
Email : niralipune@pragationline.com

➤ DISTRIBUTION CENTRES

PUNE

Nirali Prakashan : 119, Budhwar Peth, Jogeshwari Mandir Lane, Pune 411002, Maharashtra
(For orders within Pune)   Tel : (020) 2445 2044, Mobile : 9657703145
  Email : niralilocal@pragationline.com

Nirali Prakashan : S. No. 28/27, Dhayari, Near Asian College Pune 411041
(For orders outside Pune)   Tel : (020) 24690204 Fax : (020) 24690316; Mobile : 9657703143
  Email : bookorder@pragationline.com

MUMBAI

Nirali Prakashan : 385, S.V.P. Road, Rasdhara Co-op. Hsg. Society Ltd.,
Girgaum, Mumbai 400004, Maharashtra; Mobile : 9320129587
Tel : (022) 2385 6339 / 2386 9976, Fax : (022) 2386 9976
Email : niralimumbai@pragationline.com

➤ DISTRIBUTION BRANCHES

JALGAON

Nirali Prakashan : 34, V. V. Golani Market, Navi Peth, Jalgaon 425001, Maharashtra,
Tel : (0257) 222 0395, Mob : 94234 91860;
Email : niralijalgaon@pragationline.com

KOLHAPUR

Nirali Prakashan : New Mahadvar Road, Kedar Plaza, 1st Floor Opp. IDBI Bank, Kolhapur 416 012
Maharashtra. Mob : 9850046155; Email : niralikolhapur@pragationline.com

NAGPUR

Nirali Prakashan : Above Maratha Mandir, Shop No. 3, First Floor,
Rani Jhanshi Square, Sitabuldi, Nagpur 440012, Maharashtra
Tel : (0712) 254 7129; Email : niralinagpur@pragationline.com

DELHI

Nirali Prakashan : 4593/15, Basement, Agarwal Lane, Ansari Road, Daryaganj
Near Times of India Building, New Delhi 110002 Mob : 08505972553
Email : niralidelhi@pragationline.com

BENGALURU

Nirali Prakashan : Maitri Ground Floor, Jaya Apartments, No. 99, 6th Cross, 6th Main,
Malleswaram, Bengaluru 560003, Karnataka; Mob : 9449043034
Email: niralibangalore@pragationline.com

Other Branches : Hyderabad, Chennai

niralipune@pragationline.com | www.pragationline.com
Also find us on www.facebook.com/niralibooks

Preface ...

The present book entitled **"Solid State Physics"** is written as per new revised syllabus prescribed for the IIIrd Semester of T.Y.B.Sc. (Physics) from June 2015. This book is targeted mainly to the undergraduate students of Pune University, but will be found useful for the graduate students and Teachers of other universities also. The book is divided into four chapters. Each chapter begins with basic concepts containing theory, set of formulae and explanatory notes for followed by a number of solved problems. Summary of contents in topic, short, long questions and unsolved problems are also given at the end of each topic.

The problems are judiciously selected and are given topic and section-wise. The approach is straight forward and step-by step solutions are elaborately provided. More importantly the relevant formulas used for solving the problems can be located in the beginning of each chapter. There are number of diagrams for illustration.

Chapter 1 in the book is devoted to the basic concepts in Solid State Physics and various Crystal Structures. Chapter 2 is basically concerned with X-Ray Diffraction and various Characterization Techniques which includes UV-Vis, TGA and SEM. Chapter 3 is concerned with Free Electron Theory and Band Theory of Metals. Chapter 4 is basically related to Magnetism, Superconductivity and their applications.

All precautions have been taken to avoid mistakes and misprint in the book. However, it is possible that some mistakes and misprints might have passed unnoticed. Such mistakes and misprint, is brought to our notice will be thankfully acknowledged.

We are thankful to Shri Jignesh Furia and staff of Nirali publication for publishing the book in attractive look. We have a pleasure to thank Mr. Santosh Bare for the bulk of typing and Mr. Kiran Velankar for proof reading. I am indebted to Mrs. Anjali Muley for line drawings, to Ravi Walodare for designing cover page and all staff in the distribution of books network.

We are also thankful to all the Marketing Staff especially Mr. Nilesh Deshmukh and others for co-ordinating the matter well in time.

Suggestions to improve the quality of the book will be gladly accepted.

AUTHORS

Syllabus ...

1. The Crystalline State (11 L)

Lattice, Basis, Translational vectors, Primitive unit cell, Symmetry operations, Different types of lattices : 2D and 3D (Bravais lattices), Miller indices, Interplaner distances, SC, BCC and FCC structures, Packing fraction, Crystal structures : NaCl, Diamond, CsCl, ZnS, HCP, Concept of reciprocal lattice and its properties with proof. Problems.

2. X-Ray Diffraction and Other Characterization Techniques (13 L)

Introduction, Crystal as a grating, Bragg's law and Bragg's Diffraction condition, Indirect and reciprocal lattice, Ewald's construction, Experimental methods of X-ray diffraction: Laue method, Rotating crystal method, Powder (Debye Scherer) method, Analysis of cubic structure by powder method.

Characterization Techniques: Thermal gravimetric analysis (TGA), UV-visible spectroscopy, Electron microscopy (SEM), Problems.

3. Free Electron and Band Theory of Metals (13 L)

Free Electron model, Energy levels and Density of orbital in 1D and 3D, Bloch theorem (statement only), Nearly free electron model, Fermi energy, Fermi level, Hall effect, Origin of energy gap, Energy bands in solids, Effective mass of electron (with derivation), Distinction between metal, semiconductor and insulator. Problems.

4. Magnetism (11 L)

Diamagnetism, Langevin theory of Diamagnetism, Application of diamagnetic material: (Superconductor). Occurrence of Superconductivity, Critical magnetic field and Meissner effect, Paramagnetism, Langevin theory of paramagnetism, ferromagnetism, ferromagnetic domains, Hysteresis, Curie temperature. Ferromagnetism, Ferrites and their applications, Antiferromagnetism, Neel temperature, Problems.

$$\square\square\square$$

Contents ...

□□□

Chapter 1...

The Crystalline State

Contents ...

William Hallowes Miller FRS (6 April 1801 – 20 May 1880) was a Welsh mineralogist and laid the foundations o modern crystallography. According to Miller it is more useful to express orientation of planes by such indices. Miller Indices is simple method to designate a plane in crystals by three numbers (h, k, l).

1.1 Introduction

- Solid State Physics / Material Science have achieved great significance during the last few decades. This may be due to many far reaching discoveries in this area. Solid-state physics is the study of rigid matter or solids, through methods such as quantum mechanics, crystallography, electromagnetism and metallurgy. It is the largest branch of condensed matter physics. Solid-state physics includes the large-scale properties of solid materials. Thus, solid-state physics forms the theoretical basis of materials science. It also has direct applications, for example in the technology of transistors and semiconductors.

- The physical properties of solids have been common subjects of scientific inquiry for centuries, but a separate field going by the name of solid-state physics did not emerge until 1940s, in particular with the establishment of the Division of Solid State Physics (DSSP) within the American Physical Society. The DSSP catered to industrial physicists, and solid-state physics became associated with the technological applications. By the early 1960s, the DSSP was the largest division of the American Physical Society.

- Solid state physics is concerned with crystals and electrons in crystals. Study of this subject began with discovery of X-ray diffraction by crystals in early years of 20th century. Laue developed an elementary theory of diffraction of X-ray by periodic array. The work done by Laue, Friedrich and Knippling proved decisively that crystals are composed of array of atoms. The studies were extended to amorphous or non-crystalline solids, glasses and liquids. The wider field of solid state is condensed matter physics and it is now largest and most vigorous area of physics.

- The present topic deals with crystal structure and properties.

Crystalline State

- The matter in solid state possesses definite volume, definite shape and definite mass. Liquids do not have definite shape but possesses definite volume. Gases do not possess either definite volume or definite shape. The separation between the neighboring atoms in the solids and liquids is of the order of few A° while in gases, it is in the order of about 25 A° and above.

- In solid state, the atoms are closely packed and voids between them are very small. Due to small voids the particle motion is limited to vibratory motion and the magnitude of inter particle force is very large. This accounts for the large densities of solids. The definite shape and volumes of solids are attributed to the definite geometric pattern involved in the close packing of particles.

- Solid state substance can maintain its shape even it is little affected by the changes in temperature and pressure. Those solids which have regularities in atomic and molecular structure are known as **crystals**.

- There are some solids such as glasses and amorphous materials; there is no orderly arrangement of atoms or molecules. Such materials are really super-cooled liquids. The crystals are distinguished from amorphous solids by the regularities in external form and sharp melting point.

- Solids are generally classified into two groups :

  (1)　Crystalline solids (polycrystalline and single crystal) and

  (2)　Amorphous solids

- **A crystalline solid is a substance whose constituent particles possess a regular orderly arrangement** e.g. sodium chloride, diamond, quartz etc. When crystalline solid is heated, its temperature rises up to its melting point. At the melting point it liquefies suddenly. On cooling, the regular geometric shape appears again as the 'solid formed'. In polycrystalline materials, the solid is made of grains which are highly ordered crystalline regions of irregular size and orientation. Single crystal has long range order.

- **An amorphous solid is a substance whose constituent particles do not possess a regular orderly arrangement.** e.g. glass, plastics, rubber, proteins etc. In amorphous solids, there is no long range regularity. But in some cases they may possess small regions of orderly arrangement. These crystalline parts of amorphous solids are known as **crystallites**. An amorphous solid does not have a sharp melting point. It undergoes a liquification over a broad range of temperature. Amorphous solids are called as super liquids because they possess disorderly arrangement like liquids. Order in amorphous solid is limited to a few molecular distances.

- Amorphous solids are isotropic because their properties such as electrical conductivity, thermal conductivity, mechanical strength and refractive index are same in all directions. On the other hand, crystalline solids are anisotropic. The physical properties of crystalline solids are different in different directions.

1.2 Lattice　　　　　　　　　　　　　　　　　(April 16)

- **An arrangement of infinite number of imaginary points in space with each point having identical surrounding is known as crystal lattice.** In a perfect crystal, there is regular orderly arrangement of atoms. The periodicity in arrangement differs in different directions. In order to understand the periodic pattern of atomic arrangement in a crystal, it is convenient to imagine points in a space about which atoms or molecules can be located. This leads us to the concept of a space lattice.

- Three dimensional **space lattice may be defined as a finite array of points in three-dimensions in which every point has identical environment as any point in the array.**

- A space lattice provides the frame work with reference to which a crystal structure can be described.

1.3 Basis   (Oct. 16)

- A lattice is merely a uniform arrangement of points in the space. A crystal structure is formed by associating with every lattice point, a unit assembly of atoms or molecules identical in composition. This unit assembly is called basis. The crystal structure is not lattice, as a lattice is imaginary concept, but can be said as a latticed array of atoms.

- The crystal lattice is shown in Fig 1.1 (a). In this figure, the arrangements of points have been shown in which any point is completely equivalent to any other point in the structure.

- **The structural unit of a crystal is called as basis which may consist of an atom, an ion or a molecule.** In Fig. 1.1 (b), a particular arrangement of two atoms is denoted by a solid dot and a circle which represent the basis.

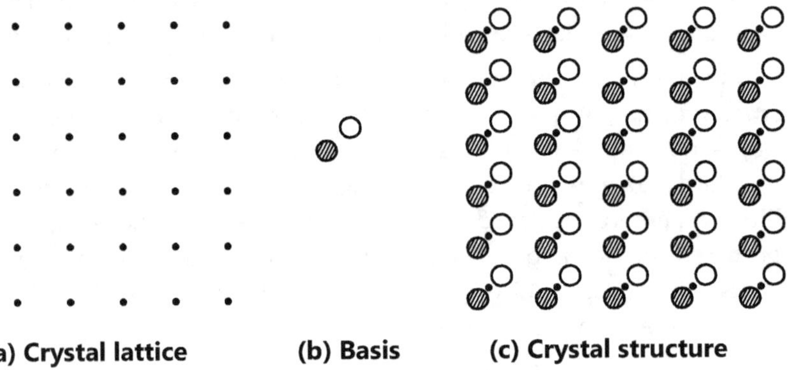

(a) Crystal lattice   **(b) Basis**   **(c) Crystal structure**

Fig. 1.1 : Crystal structure

- When the basis is attached to every point of the lattice [as shown in Fig. 1.1 (a)] crystal structure is formed as shown in Fig. 1.1 (c).

Thus,

Lattice + Basis = Crystal structure

1.4 Translational Vectors

- Consider a two-dimensional square array of points as shown in Fig. 1.2. Let the distance between two successive points be **a** and **b** along X-axis and Y-axis respectively. By repeated translation of two vectors $\vec{a}$ and $\vec{b}$ on the plane of the paper, we can generate the square array. The vectors $\vec{a}$ and $\vec{b}$ are known as **fundamental translation vectors**. These vectors generate square array. If we locate ourselves at any point in the array and look out in a particular direction, the arrangement and orientation of the lattice points are same, irrespective of our position.

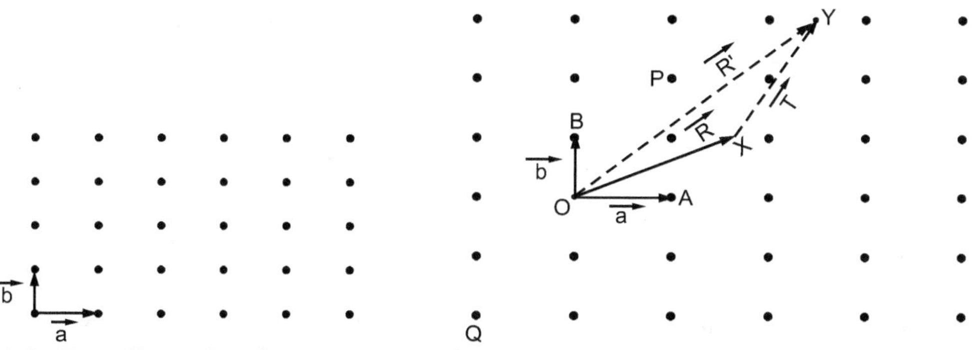

Fig. 1.2 : Two dimensional square array of points   Fig. 1.3 : Translational vectors

- The space lattice can be defined with the help of translation operation. In order to understand the translation operation, consider the case of small group of lattice points in two-dimensional space as shown in Fig. 1.3.

- Let us choose the point O as an origin. Consider the lattice points A and B in Fig. 1.3. Let $\overrightarrow{OA} = \overrightarrow{a}$ and $\overrightarrow{OB} = \overrightarrow{b}$. All other lattice points can be obtained with respect to origin O by a vector $\overrightarrow{T}$ such that $\overrightarrow{T} = n_1 \overrightarrow{a} + n_2 \overrightarrow{b}$

  where n_1 and n_2 are integers and $\overrightarrow{a}$ and $\overrightarrow{b}$ are translational vectors.

  For point P, $n_1 = 1$ and $n_2 = 2$. Therefore, $\overrightarrow{T} = \overrightarrow{a} + 2\overrightarrow{b}$.

  For point Q, $n_1 = -1$ and $n_2 = -2$. Therefore, $\overrightarrow{T} = -\overrightarrow{a} - 2\overrightarrow{b}$.

- In a similar manner, all other lattice points are obtained with respect to the origin O by translation operation. If the starting point is a lattice point, then upon translation we reach other lattice points and thus span entire lattice structure.

- We can start from any point not necessarily be a lattice point such as a point X. Let $\overrightarrow{OX} = \overrightarrow{R}$. The point Y situated at $\overrightarrow{R'}$ with respect to origin O can be obtained from $\overrightarrow{R}$ using the translation operation.

$$\overrightarrow{R'} = \overrightarrow{R} + \overrightarrow{T} = \overrightarrow{R} + \overrightarrow{a} + 2\overrightarrow{b}$$

where,
$$\overrightarrow{T} = \overrightarrow{a} + 2\overrightarrow{b}$$

- In the present case the points situated at $\overrightarrow{R}$ and $\overrightarrow{R'}$ have identical orientation with respect to the neighbouring lattice points.

- In going from $\overrightarrow{R}$ to $\overrightarrow{R'}$, we translate ourselves by a vector $\overrightarrow{T} = n_1 \overrightarrow{a} + n_2 \overrightarrow{b}$, where n_1 and n_2 are integers. Hence, $\overrightarrow{a}$ and $\overrightarrow{b}$ are called translation vectors. The translation vectors $\overrightarrow{a}$ and $\overrightarrow{b}$ can be chosen in a number of ways. Two such ways are shown in Fig. 1.4, where $\overrightarrow{a_1}, \overrightarrow{b_1}$ and $\overrightarrow{a_2}, \overrightarrow{b_2}$ are two sets of translation vectors.

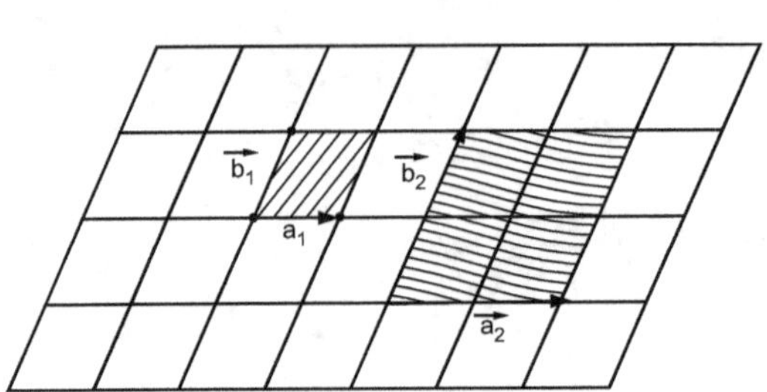

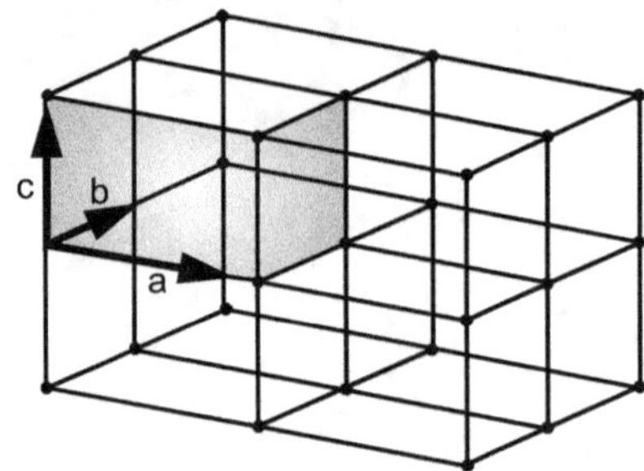

Fig. 1.4 : Translation vectors   **Fig. 1.5 : Three-dimensional space lattice**

- In Fig. 1.4, the translation vectors $\vec{a_1}, \vec{b_1}$ or $\vec{a_2}, \vec{b_2}$ form two adjacent sides of the parallelogram.

- When we complete the parallelogram using $\vec{a_1}, \vec{b_1}$ or $\vec{a_2}, \vec{b_2}$, we get a structure which can be supposed to span the entire lattice structure by periodically repeating it throughout.

- Thus, there are several choices of translation vectors. However, **there exists a pair of translation vectors such that the elementary parallelogram formed by them is of minimum area. Such translation vectors are called primitive or fundamental translation vectors.**

- In Fig. 1.4, $\vec{a_1}, \vec{b_1}$ are primitive translation vectors, whereas $\vec{a_2}, \vec{b_2}$ are non-primitive translation vectors. The translation vectors for which the translation operation contains integral coefficients are called primitive translation vectors. The translation operation T is given by
$$\vec{T} = \text{(integer)}\ \vec{a} + \text{(integer)}\ \vec{b}$$
where $\vec{a}$ and $\vec{b}$ are called primitive translation vectors.

- On the other hand, the translation vectors for which the translation operation contains non-integral coefficients are called **non-primitive translation** vectors. Above discussion can be extended for a three-dimensional space lattice (as shown in Fig. 1.5).

For a three-dimensional lattice, the translation operation is given by
$$\vec{T} = n_1 \vec{a} + n_2 \vec{b} + n_3 \vec{c}$$
where n_1, n_2 and n_3 are integers and $\vec{a}, \vec{b}, \vec{c}$ are called fundamental translation vectors.

1.5 Primitive Unit Cell

- Consider a part of two-dimensional lattice as shown in Fig. 1.6. The parallelograms formed by the translation vectors may be used as building blocks for constructing the complete lattice and are known as unit cells of the lattice.

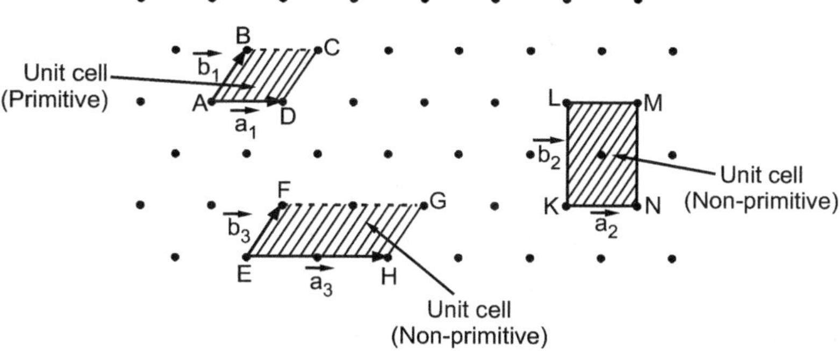

Fig. 1.6 : Primitive unit cell

- There are many ways in which a cell can be drawn in a given space lattice. Three such possibilities are shown in Fig. 1.6. The unit cell ABCD is formed by translation vectors $\vec{a_1}$, $\vec{b_1}$, the unit cell KLMN is formed by translation vectors $\vec{a_2}$, $\vec{b_2}$ and so on. The choice of unit cell is not unique. To construct unit cell primitive or non-primitive translation vectors can be used.

- **If primitive translation vectors are used to construct a unit cell then it is called primitive unit cell.** *On the other hand, if non-primitive translation vectors are used to construct a unit cell, the cell is non-primitive unit cell.*

- In Fig. 1.6, the parallelogram ABCD represents a two-dimensional primitive cell, whereas EFGH and KLMN are non-primitive cells. *The primitive cell is of minimum area and all the lattice points belonging to it lie at its corner only.* Therefore, the effective number of lattice points in a primitive unit cell is one.

- For a three-dimensional lattice, the unit cells are of the form of parallelopiped. The primitive unit cell in three-dimensional space is formed by primitive translation vectors $\vec{a}$, $\vec{b}$ and $\vec{c}$, which is parallelopiped as shown in Fig. 1.7. It has the volume $V_c = \vec{a} \cdot (\vec{b} \times \vec{c})$. The volume of primitive unit cell is $V_c = \vec{a} \cdot (\vec{b} \times \vec{c})$ and is the cell having smallest volume. **Thus, the unit cell whose concurrent sides qualify to be the primitive translation vectors is called primitive cell.**

- In this cell, the lattice points lie at its corner only. Therefore, the effective number of lattice points in a primitive cell is one. A non-primitive cell may have the lattice points at the corners as well as at other locations such as inside and/or on the surface of the cell. Therefore, the effective number of lattice points in a non-primitive cell is greater than one.

- The small identical building units, which are repeated again in a three-dimensional space building the crystalline structure, are called unit cells.

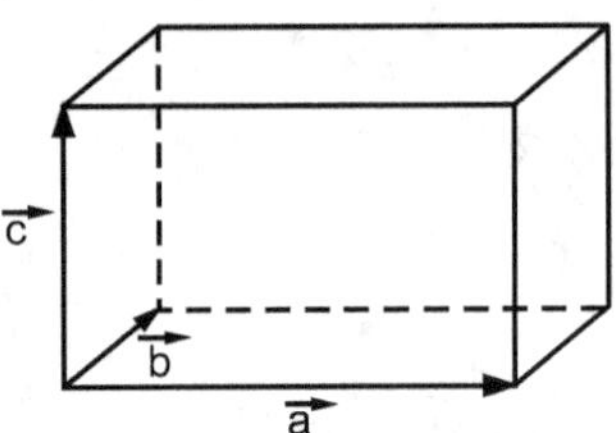

Fig. 1.7 : Primitive unit cell

- The unit cell may also be regarded as the basic building block that makes up the crystal, each one indistinguishable from the next. The unit cell may be primitive or non-primitive unit cell.

1.6 Symmetry Operations (April 17)

- To understand the concept of symmetry operation, consider the example of an equilateral triangle ABC with identical atoms situated at its vertices [Refer Fig. 1.8 (a)]. If the triangle is rotated by 120° about an axis perpendicular to the plane and passing through centroid O, the triangle would assume an orientation as shown in Fig. 1.8 (b). As the atoms at the vertices are identical in all respect, the new structure is the same as the earlier one. Thus, the operation of rotation that we have performed on the triangle ABC has carried the structure into itself.

- Now consider another example of a plane thin mirror held perpendicular to the plane of triangle ABC and containing the altitude BP as shown in Fig. 1.8 (a). Due to reflection, we get structures which exactly coincide with initial one. Hence mirror reflection symmetry exists about the plane containing BP.

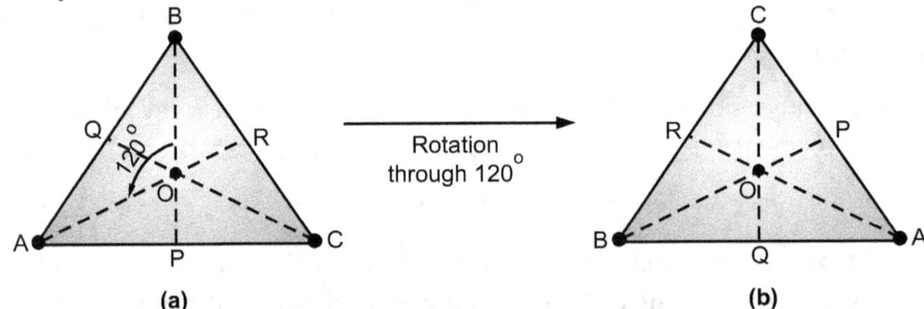

Fig. 1.8 : Equilateral triangle ABC

- **Thus, there exist certain operations which when operated on a given structure, the structure remains unchanged or invariant. Such operations are called symmetry operations and the structure is said to possess symmetry under that operation.**

Different symmetry operations :

- Symmetry operations carried out about a point or a line are called **point group symmetry** operations. The symmetry operations performed by translations as well as rotation are termed as **space group symmetry**. Crystal possesses both types of symmetries.

- Following are the types of symmetry operations that commonly occur :

  (i)   Translational symmetry.

  (ii)   Rotational symmetry.

  (iii)   Symmetry under mirror reflection.

  (iv)   Inversion through a point or centre of symmetry.

  (v)   Screw symmetry.

  (vi)   Glide plane symmetry.

- The rotation, reflection and inversion are point operations. When point group symmetry operations are combined with translational symmetry operations then the operations become space group symmetry operations.

  (i) Translational symmetry : We can translate the entire crystal structure by translation operation. The lattice point at $\vec{r}$, under translation vector operation $\vec{T}$, gives another point $\vec{r'}$ which is identical to $\vec{r}$ i.e. $\vec{r'} = \vec{r} + \vec{T}$ and $\vec{T} = n_1 \vec{a} + n_2 \vec{b} + n_3 \vec{c}$ where n_1, n_2 and n_3 are integers and $\vec{a}$, $\vec{b}$ and $\vec{c}$ are fundamental translation vectors.

  (ii) Rotational symmetry : To understand the term rotational symmetry, consider an equilateral triangle ABC [Refer Fig. 1.8 (a)]. If this triangle is rotated through 360 (i.e. one complete rotation), it is carried out into itself (i.e. coincides with initial one) three times at positions 120°, 240° and 360° rotation.

- The minimum angle of rotation which leaves the structure unchanged in this case is 120°.

- The fold number is given by the formula

$$\text{Fold number} = \frac{2\pi}{\substack{\text{Minimum angle (non-zero) of rotation} \\ \text{which leaves the structure unchanged}}}$$

- Thus, equilateral triangle possesses a 3-fold rotational axis of symmetry around the rotational axis.

- When a cube is rotated about the vertical axis through 360°, there are four positions of cube which are coincident with its original position as shown in Fig. 1.9 (a). When the cube is rotated through 90°, the new position is completely identical original position. This rotation axis is known as four-fold rotation axis or four-fold axis of symmetry.

- Similarly, when a cube is rotated about a solid diagonal through 120°, the new position is exactly same as original position or self-coincidence. Therefore, such axis of rotation is known as three-fold rotation axis. It is shown in Fig. 1.9 (b).

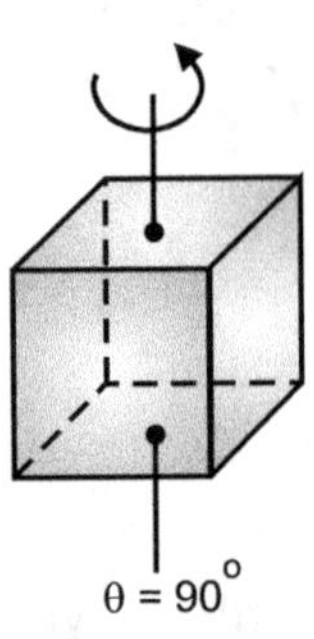
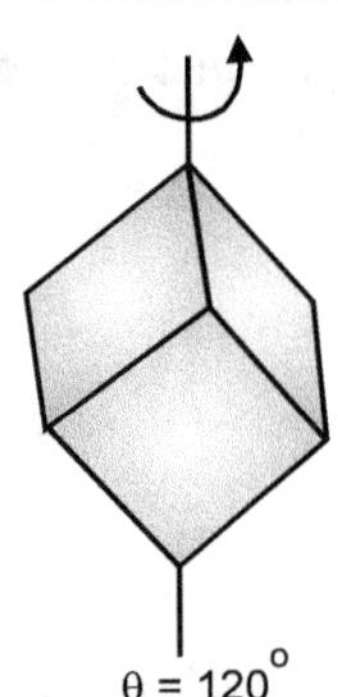
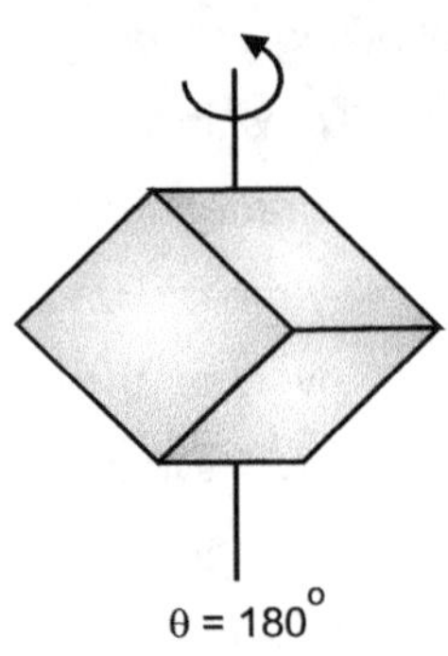

$\theta = 90°$ $\theta = 120°$ $\theta = 180°$

(a) Four-fold rotation axis   **(b) Three-fold rotation axis**   **(c) Two-fold rotation axis**

Fig. 1.9

- Consider the axis of rotation passing through the middle points of a pair of opposite parallel edges as shown in Fig. 1.9 (c).

- When a cube is rotated through 180° about this axis, the new position of cube is exactly same as original position. Therefore, such axis of rotation is known as two-fold rotation axis. It is found that crystalline solids show only 1, 2, 3, 4 and 6 fold symmetry. They do not show 5-fold symmetry.

   (iii) Symmetry under mirror reflection : A structure is said to have reflection symmetry, if it is left invariant or unaltered in everyway after being reflected by plane.

- Consider equilateral triangle ABC as shown in Fig. 1.10. If we fold the figure about the plane BO, the two parts become coincident. This type of coincidence can be brought by a plane mirror. If the plane mirror is held perpendicular to the plane of triangle and containing the altitude BO the image reproduce the other half portion of the triangle. Such a plane is called a plane of symmetry.

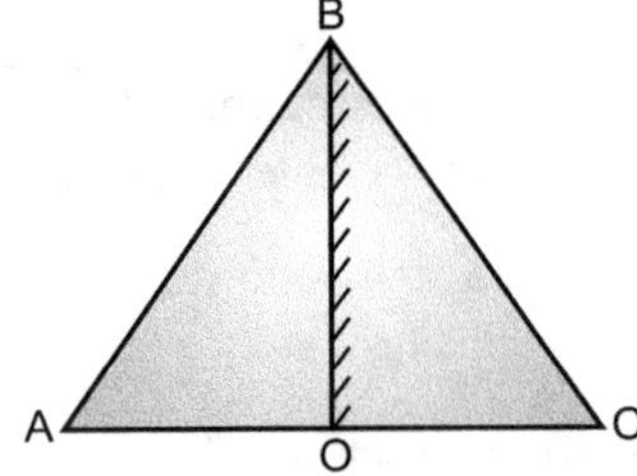

Fig. 1.10 : Equilateral triangle ABC

- A cubical crystal can be cut into two equal halves such that one half is the reflection of the other half. There is a certain plane along which, if crystal is cut and then placed on mirror, the image reproduce the other half of the crystal. Such a plane is called a plane of symmetry. In Fig. 1.11 in a cube, there are three planes of symmetry parallel to the faces of cube are shown.

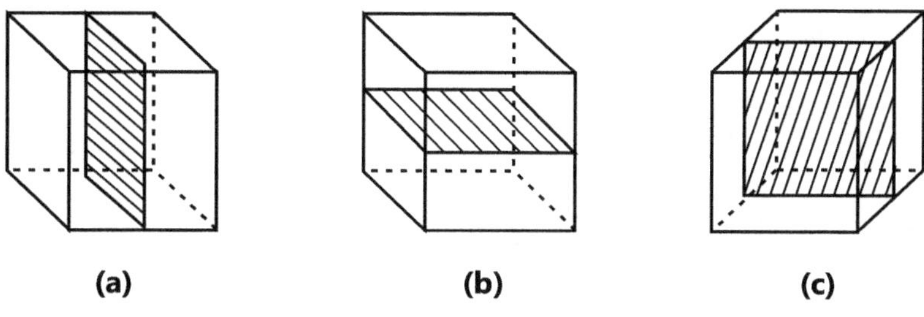

(a) **(b)** **(c)**

Fig. 1.11 : Planes of symmetry

(iv) Inversion through a point or Centre of symmetry : The centre of symmetry is also known as the centre of inversion (I). It is the point in the crystal such that if a line is drawn from any point on the crystal through this point and produces an equal distance on the other side of this centre, it meets an identical point.

• Consider a point A on circle as shown in Fig. 1.12 (a). A line AO is drawn and produced it to B such that AO = OB. The point B lies in opposite quadrant on the circumference of circle. So the point O is the centre of symmetry.

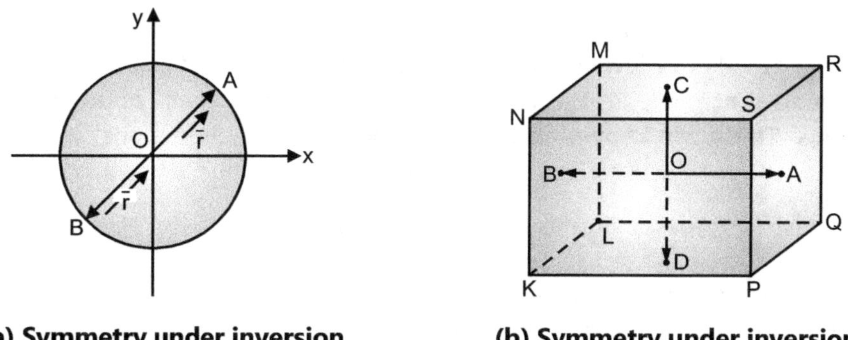

(a) Symmetry under inversion **(b) Symmetry under inversion**

Fig. 1.12

• The same is true for a rectangular parallelopiped about a point O in Fig. 1.12 (b). Consider a point A on the plane PQRS. A line AO is drawn and produced to B such that AO = OB. The point B lies on opposite face KLMN. So the point O is called centre of symmetry.

(v) Screw symmetry : Symmetry operation due to compound symmetry operation namely translation and rotation applied successively is called as screw operation.

• Consider lattice points situated on the surface of imaginary cylinder as shown in Fig. 1.13. If we rotate the cylinder about its axis through 180° and then translate through a distance r parallel to the axis of cylinder, we see that the structure merges into itself. As the above operation is just like screw, the structure is said to possess screw symmetry.

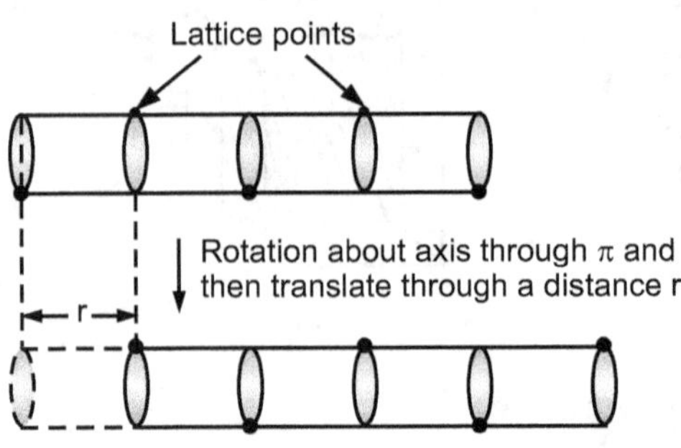

Fig. 1.13 : Screw symmetry

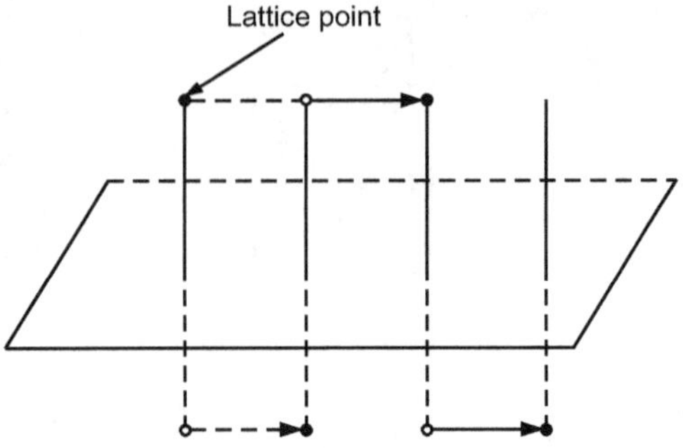

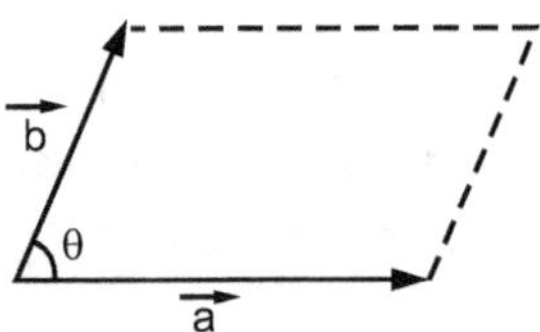

Fig. 1.14 : Glide plane symmetry

(vi) Glide plane symmetry : When a mirror plane is combined with a simultaneous translational operation in crystal, one gets glide plane. Glide plane in the crystal is always parallel to mirror plane. This type of symmetry is shown in Fig. 1.14. The compound operation leaves the structure invariant or unchanged. The structure possessing this property is said to have glide plane symmetry.

1.7 Different Types of Lattices

(a) Bravais lattices in two dimensions : It was shown by Bravais that only **14 different networks of lattices can be generated in which lattice points arranged in 3-dimensional space of every network are such that each point has identical surroundings. These are known as Bravais lattices.** In two-dimensional space, there are five different lattices which can be generated using translation of two coplanar vectors $\vec{a}$ and $\vec{b}$. These five types of Bravais lattices are as follows :

(i) Oblique lattice : In this type, $\vec{a} \neq \vec{b}$ and angle between $\vec{a}$ and $\vec{b}$ is different from 90° or 120°. Oblique lattice is as shown in Fig. 1.15. The conventional unit cell is a parallelogram.

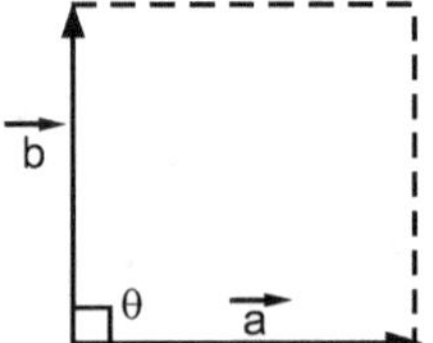

Fig. 1.15 : Oblique lattice

($|\vec{a}| \neq |\vec{b}|$, $\theta \neq 90°$)

Fig. 1.16 : Square lattice

($|\vec{a}| = |\vec{b}|$, $\theta = 90°$)

(ii) Square lattice : In this type, $|\vec{a}| = |\vec{b}|$ and $\theta = \dfrac{\pi}{2}$. The conventional unit cell is a square.

(iii) Hexagonal lattice : In this type, $|\vec{a}| = |\vec{b}|$, and $\theta = \dfrac{2\pi}{3}$ or 120°. In hexagonal lattice, a regular hexagon is formed when all the nearby lattice points about a given lattice point O are connected as indicated in Fig. 1.17. The primitive cell is a rhombus of 120° angle.

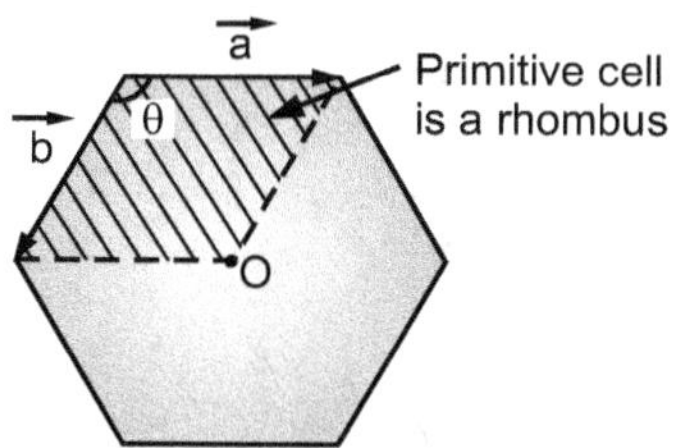

Fig. 1.17 : Hexagonal lattice

$|\vec{a}| = |\vec{b}|$ **and** θ **= 120°**

(iv) Rectangular lattice : In this type of lattice, $|\vec{a}| \neq |\vec{b}|$ but $\theta = \dfrac{\pi}{2}$. The conventional unit cell is a rectangle.

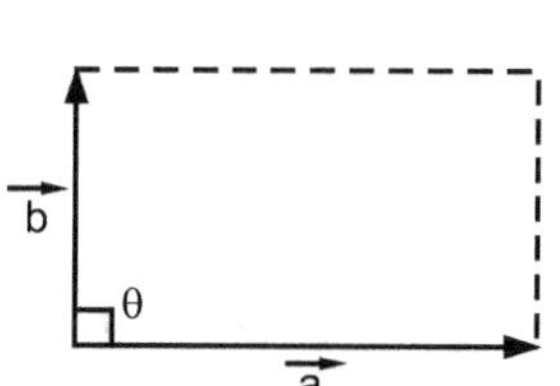

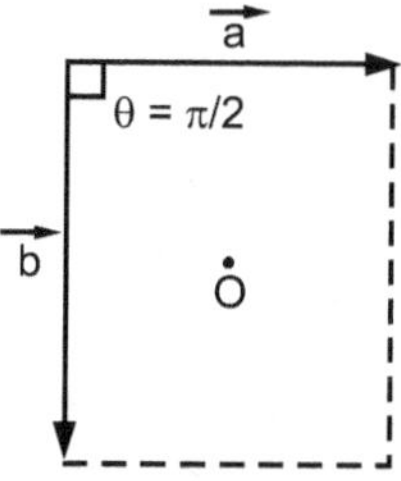

Fig. 1.18 : Rectangular lattice   **Fig. 1.19 : Centered rectangular lattice**

$|\vec{a}| \neq |\vec{b}|$ **but** $\theta = \pi/2$ $|\vec{a}| \neq |\vec{b}|$ **but** $\theta = \pi/2$

(v) Centred rectangular lattice : In this type of lattice, $|\vec{a}| \neq |\vec{b}|$ but $\theta = \pi/2$. A rectangle is formed when all the nearby lattice points about a given lattice point O are connected. (Refer Fig. 1.19).

(b) Three-dimensional lattices : Fourteen different networks of lattices can be generated in three-dimensional space by repeated translation of three non-coplanar vectors $\vec{a}$, $\vec{b}$, $\vec{c}$. These 14 Bravais lattice types are grouped into seven types of conventional unit cell. The angles α, β and γ represent the angle between the vectors and $\vec{a}$ and $\vec{b}$ respectively as shown in Fig. 1.20. The lengths $|\vec{a}|$, $|\vec{b}|$, $|\vec{c}|$ and angles α, β, γ are collectively known as lattice parameters or lattice constants of a unit cell.

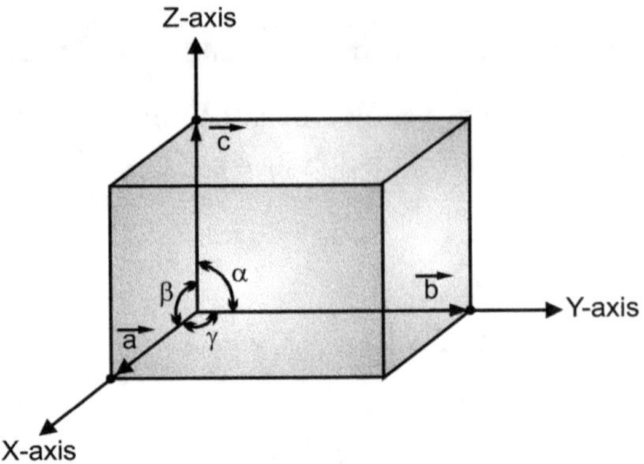

Fig. 1.20 : Three-dimensional lattices

All the crystals may be classified into following seven systems :

(1)   Cubic

(2)   Tetragonal

(3)   Orthorhombic

(4)   Rhombohedral

(5)   Hexagonal

(6)   Monoclinic

(7)   Triclinic

The shapes of their unit cells are shown in Fig. 1.21 and their lattice parameters are summarized in Table 1.1.

(1)   Cubic structure : In this crystal structure, all the three lengths of the unit cell are equal and are at right angle to each other. Therefore,

$$a = b = c \text{ and } \alpha = \beta = \gamma = 90°$$

(2)   Tetragonal structure : In this crystal arrangement, three axes are at right angles. Two sides are equal while third side is different in length. Therefore,

$$a = b \neq c$$

and $\qquad \alpha = \beta = \gamma = 90°$

(3)   Orthorhombic structure : In this crystal arrangement, three unequal sides are at right angles. Therefore,

$$a \neq b \neq c$$

and $\qquad \alpha = \beta = \gamma = 90°$

(4)   Rhombohedral structure : In this crystal arrangement, three equal sides are equally inclined but at an angle other than right angle. Therefore,

$$a = b = c$$

and $\qquad \alpha = \beta = \gamma \neq 90°$

(5) Hexagonal structure : In this crystal arrangement, three axes of the unit cell are equal in one plane at 120° from each other and a fourth axis normal to this plane. The interval along the fourth axis is unique. Therefore,

$$a = b \neq c$$

and

$$\alpha = \beta = 90°$$

$$\gamma = 120°$$

(i) Cubic :

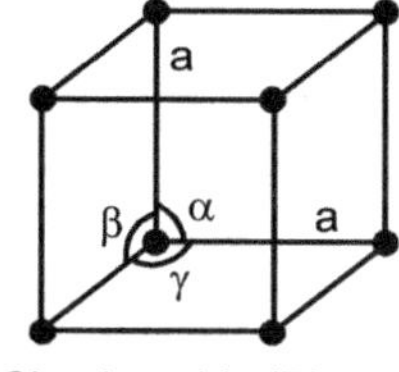

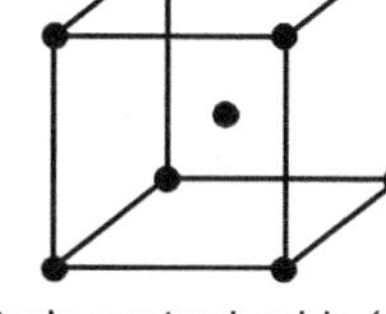

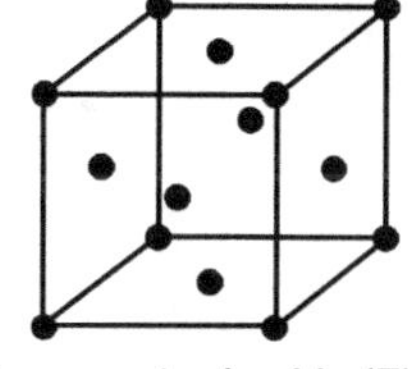

Simple cubic (P)   Body centred cubic (J)   Face-centred cubic (F)

(ii) Tetragonal :

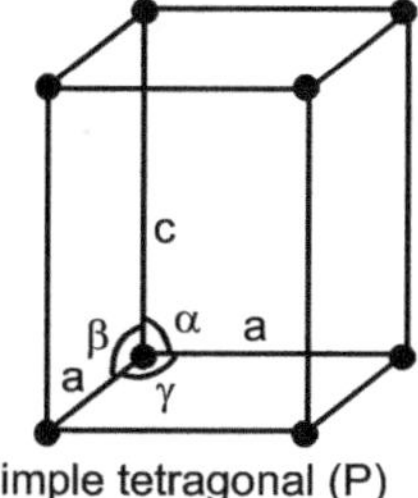

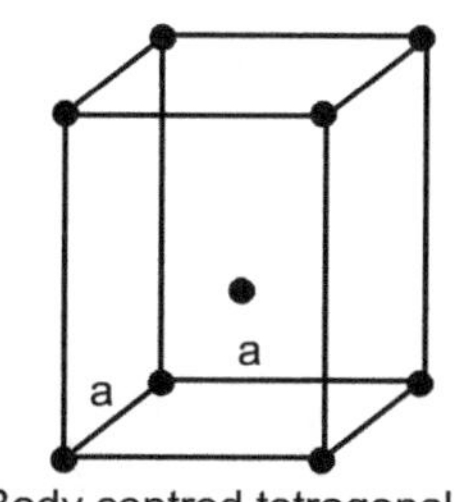

Simple tetragonal (P)   Body centred tetragonal (I)

(iii) Orthorhombic :

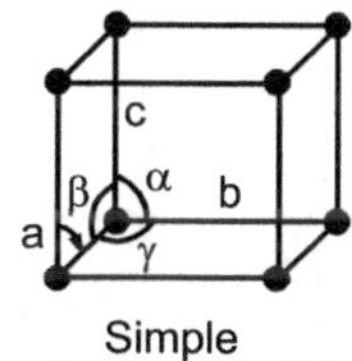

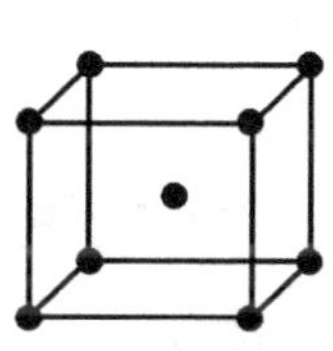

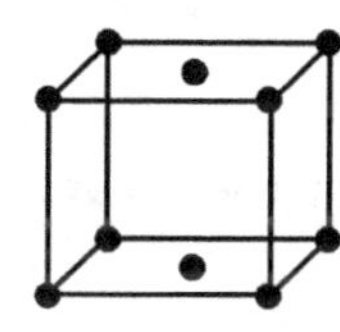

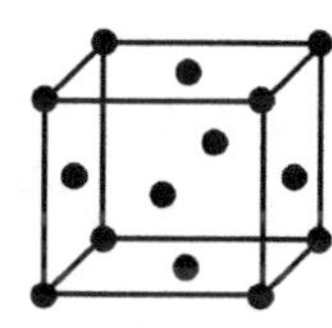

Simple orthorhombic (P)   Body centred orthorhombic (I)   End centred orthorhombic (C)   Face-centred orthorhombic (F)

(iv) Rhombohedral or Trigonal :

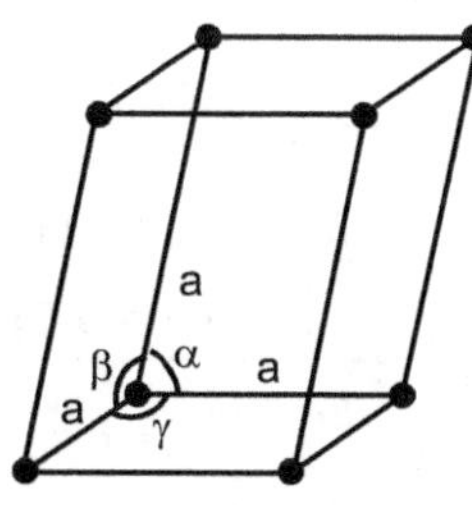

Simple rhombohedral (P)

(v) Hexagonal :

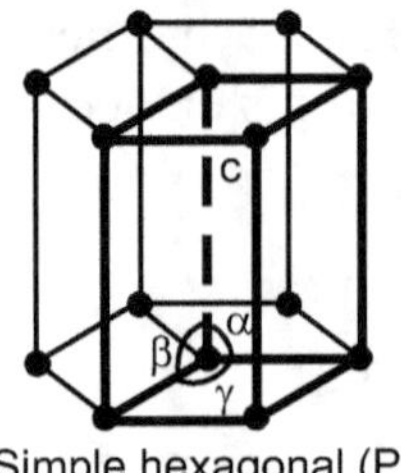

Simple hexagonal (P)

(vi) Monoclinic :

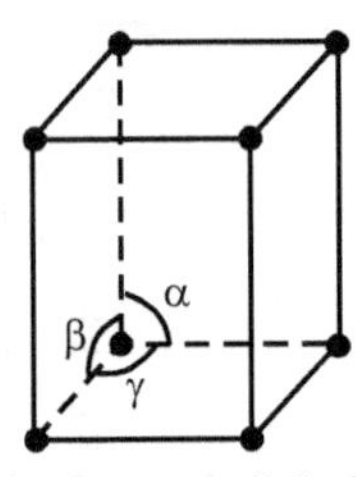

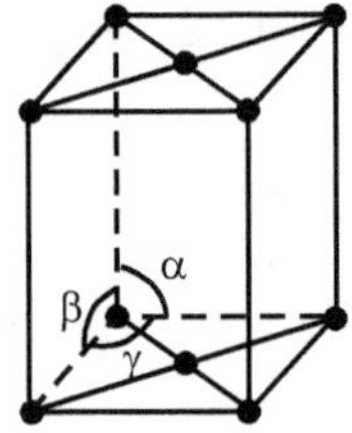

Simple monoclinic (P)   Base-centred monoclinic (C)

(vii) Triclinic :

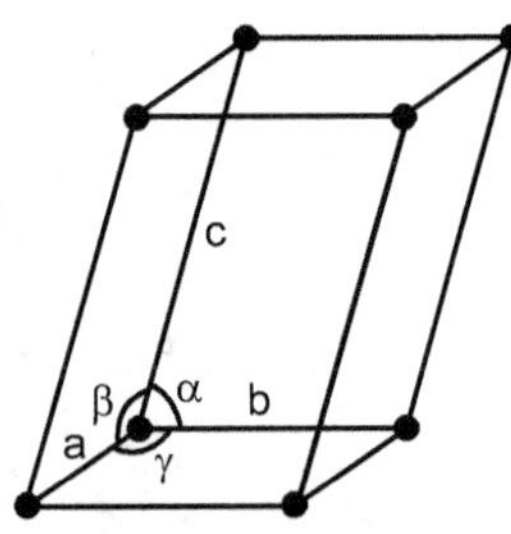

Simple triclinic (P)

Fig. 1.21 : The Bravais lattices in three dimensions

Table 1.1

Sr. No.	Crystal system	Lattice parameters	Bravais lattice	Number of different lattices in the system and symbol	Examples
1.	Cubic	$a = b = c$ $\alpha = \beta = \gamma = 90°$	Simple (SC) Body-centred (BCC) Face-centred (FCC)	3 (P, I, F)	Cu, Ag, Fe, NaCl, CsCl etc.
2.	Tetragonal	$a = b \neq c$ $\alpha = \beta = \gamma = 90°$	Simple Body-centred	2 (P, I)	B-Sn, TiO_2, SnO_2

... Contd.

3.	Orthorhombic	$a \neq b \neq c$ $\alpha = \beta = \gamma = 90°$	Simple Body-centred Base or end-centred Face-centred	4 (P, I, C, F)	Ga, Fe_3C, KNO_3 etc. (cementite)
4.	Rhombohedral to Trigonal	$a = b = c$ $\alpha = \beta = \gamma \neq 90°$	Simple	1 (P)	As, Sb, $CaSO_4$ etc.
5.	Hexagonal	$a = b \neq c$ $\alpha = \beta = 90°,$ $\gamma = 120°$	Simple	1 (P)	Mg, Zn, Cd, SiO_2 etc.
6.	Monoclinic	$a \neq b \neq c$ $a = \gamma = 90° \neq \beta$	Simple Base-centred	2 (P, C)	Na_2SO_4, $FeSO_4$ etc.
7.	Triclinic	$a \neq b \neq c$ $\alpha \neq \beta \neq \gamma \neq 90°$	Simple	1 (P)	$K_2Cr_2O_7$, $CuSO_4$ etc.

(6)   Monoclinic structure : In this crystal arrangement, three sides of unit cell are of different lengths. One of the axes is at right angles to the other two axes, but the other two axes are not at right angles to each other. Therefore,

$$a \neq b \neq c$$
$$\alpha = \beta = 90° \neq \gamma$$

(7)   Triclinic structure : In this crystal arrangement, three unequal sides are unequally inclined and none being at right angles. Therefore,

$$a \neq b \neq c$$
and
$$\alpha \neq \beta \neq \gamma \neq 90°$$

1.8 Miller Indices (Oct. 17, 15)

* Orientations of planes or faces in a crystal may be described in terms of their intercepts on the three axes.

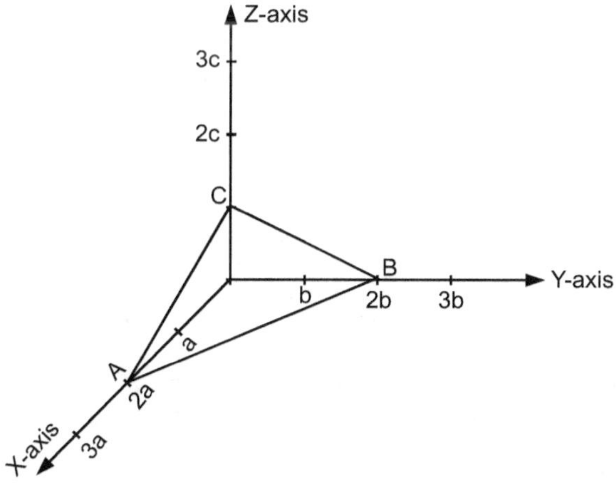

Fig. 1.22 : Miller Indices

- Consider a plane ABC which has intercepts 2-axial units on X-axis, 2-axial units on Y-axis and one axial unit on Z-axis as shown in Fig. 1.22. The numerical parameters of the faces are 2, 2 and 1. Therefore, its orientation is (2, 2, 1). However, according to Miller, **it is more useful to describe the orientation of a plane by the reciprocal of its numerical parameters rather than by its linear parameter. Such parameters are called Miller Indices.**

Procedure for finding Miller indices :

(i) First find the intercepts x, y and z of the plane on the three axes in terms of lattice constants a, b, c.

(ii) Express them in terms of axial units. Let $x = p_1a$, $y = p_2b$ and $z = p_3c$.

(iii) Take the reciprocal of the number, p_1, p_2 and p_3 i.e. we get $\dfrac{1}{p_1}$, $\dfrac{1}{p_2}$, $\dfrac{1}{p_3}$.

(iv) Convert these reciprocals into whole numbers by multiplying each with their LCM (L) and bracket the result in a parenthesis.

$$\therefore\ h = \frac{L}{p_1},\ k = \frac{L}{p_2}\ \text{and}\ l = \frac{L}{p_3}.\ \text{Here } h,\ k,\ l \text{ are called Miller indices.}$$

The steps in the determination of Miller indices of a crystal plane are illustrated with the help of Fig. 1.22.

(i) Find the intercepts on the axes in terms of the lattice constants a, b, c.

	x	y	z
	2a	2b	c
	p_1a	p_2b	p_3c
$\therefore$	$p_1 = 2$	$p_2 = 2$	$p_3 = 1$

(ii) Take the reciprocals of the numbers p_1, p_2 and p_3.

$$\frac{1}{2} \qquad \frac{1}{2} \qquad \frac{1}{1}$$

(iii) Convert these reciprocals into whole numbers by multiplying each with their LCM.

$$h = 2 \times \frac{1}{2} = 1,\ k = 2 \times \frac{1}{2} = 1,\ l = 2 \times \frac{1}{1} = 2$$

(iv) Enclose them in bracket.

$$(1\ 1\ 2)$$

Thus, Miller indices of the plane (as shown in Fig. 1.22) are (1 1 2).

Miller Indices of Cubic Crystal Planes : (Oct. 15)

While finding Miller indices of a plane, we have to consider the following points :

(1) When a plane is parallel to one of the co-ordinate axes, it is said to meet that axis at infinity. As $\dfrac{1}{\infty}$ = 0, therefore, the Miller index for that axis is zero.

(2) When the intercept of a plane is on the negative part of any axis, the Miller index is distinguished by a minus sign placed directly over it.

Consider the shaded planes in Fig. 1.23.

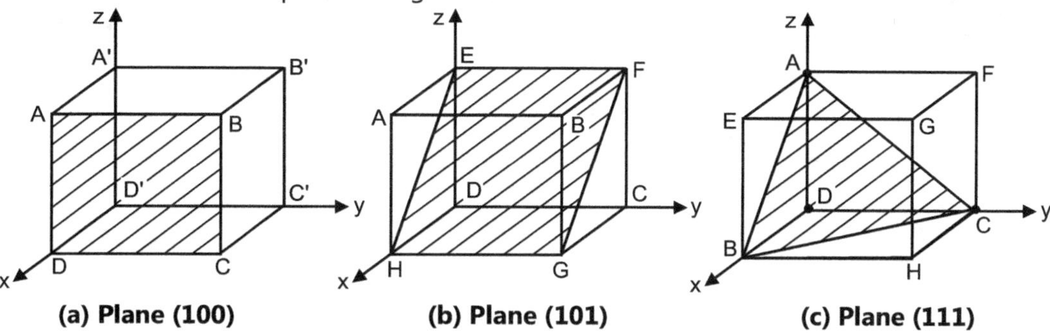

(a) Plane (100)   **(b) Plane (101)**   **(c) Plane (111)**

Fig. 1.23 : Miller indices of planes

- The plane ABCD in Fig. 1.23 (a) whose intercepts are 1a, ∞a, ∞a. Its numerical parameters are 1, ∞ and ∞. Hence, the reciprocals of numerical parameters give Miller indices

$$\left(\frac{1}{1}:\frac{1}{\infty}:\frac{1}{\infty}\right) \quad \text{or } (1\,0\,0)$$

- Now, consider the shaded plane EFGH of Fig. 1.23 (b). It has intercepts unity on x-axis and z-axis but parallel to y-axis. The intercepts of the plane EFGH in axial units are 1a, ∞b and 1c. Its numerical parameters are 1, ∞ and 1. By taking reciprocals of numerical parameters, we get Miller indices. Therefore, the Miller indices of the plane EFGH are (101).

- Similarly, for the shaded plane ABC in Fig. 1.23 (c) which are equal intercepts on the three axes 1a, 1a, 1a. Therefore, its numerical parameters are 1, 1 and 1. By taking reciprocal of numerical parameters, we can obtain Miller indices. Therefore,

$$\left(\frac{1}{1}:\frac{1}{1}:\frac{1}{1}\right) \quad \text{gives (111)}$$

1.9 Interplaner Distances (Oct. 17, April 16)

- **Interplaner distance is the spacing between two parallel planes in a given cell :** To obtain the formula for interplaner distance, consider a simple unit cell in which the co-ordinate axes are perpendicular to each other. Fig. 1.24 (a) shows the unit cell of a crystal. Consider a plane ABC having Miller indices (hk*l*). This plane belongs to family of planes whose Miller indices are (hk*l*) because Miller indices represent a set of paralle planes.

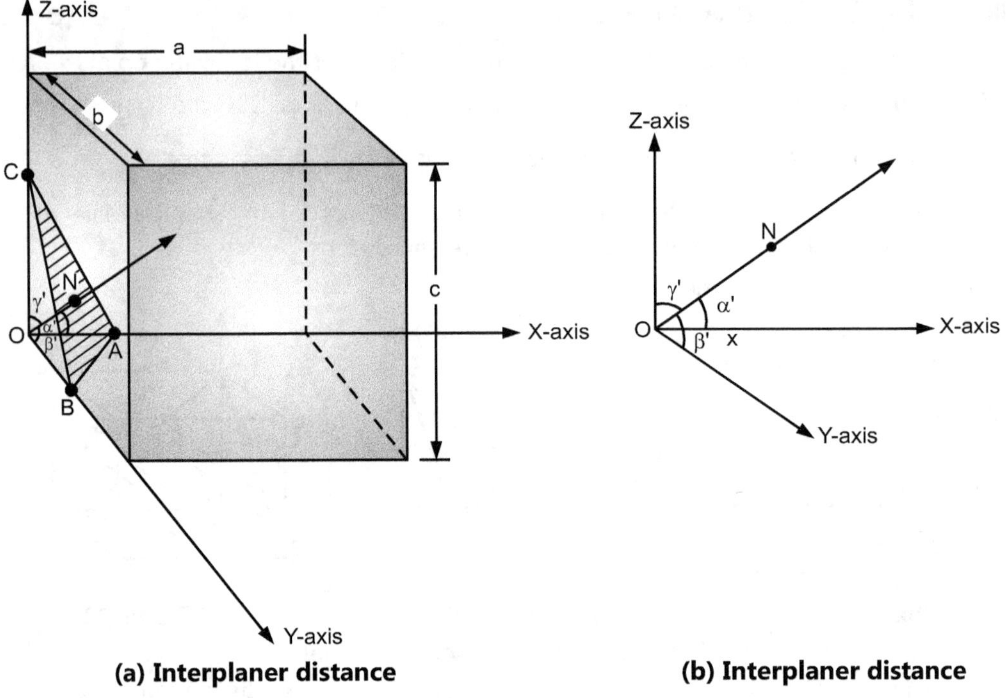

(a) Interplaner distance　　　　**(b) Interplaner distance**

Fig. 1.24

The distance ON is the interplaner distance of the family of planes. Let it be 'd'. Let α', β and γ' be the angles between co-ordinate axes X, Y and Z and ON respectively.

The intercepts of the plane ABC on the three axes are

$$OA = \frac{a}{h}, \quad OB = \frac{b}{k} \quad \text{and} \quad OC = \frac{c}{l} \qquad \qquad \text{... (1.1)}$$

As ON is normal to the plane ABC, we write

$$\cos \alpha' = \frac{d}{OA}, \quad \cos \beta' = \frac{d}{OB} \quad \text{and} \quad \cos \gamma' = \frac{d}{OC} \qquad \text{... (1.2)}$$

Using equation (1.1) in equation (1.2), we get

$$\cos \alpha' = \frac{dh}{a}, \quad \cos \beta' = \frac{dk}{b} \quad \text{and} \quad \cos \gamma' = \frac{dl}{c}$$

According to the law of direction cosines

$$\cos^2 \alpha' + \cos^2 \beta' + \cos^2 \gamma' = 1, \text{ we get}$$

$$\frac{d^2 h^2}{a^2} + \frac{d^2 k^2}{b^2} + \frac{d^2 l^2}{c^2} = 1$$

or

$$d^2 \left(\frac{h^2}{a^2} + \frac{k^2}{b^2} + \frac{l^2}{c^2} \right) = 1$$

or
$$\frac{1}{d^2} = \frac{h^2}{a^2} + \frac{k^2}{b^2} + \frac{l^2}{c^2}$$

or
$$d = \frac{1}{\sqrt{\dfrac{h^2}{a^2} + \dfrac{k^2}{b^2} + \dfrac{l^2}{c^2}}} \qquad \qquad \text{... (1.3)}$$

The expression given by equation (1.3) is known as expression for interplaner distance.

Case (1) : For cubic system,   **(April 16)**

$$a = b = c$$

$\therefore$
$$d^2\left[\frac{h^2}{a^2} + \frac{k^2}{a^2} + \frac{l^2}{a^2}\right] = 1$$

or
$$\frac{d^2}{a^2}(h^2 + k^2 + l^2) = 1$$

$\therefore$
$$d^2 = \frac{a^2}{h^2 + k^2 + l^2}$$

or
$$d_{khl} = \frac{a}{\sqrt{h^2 + k^2 + l^2}}$$

Case (2) : For tetragonal system,

$$a = b$$

Therefore, above equation gives,

$$d^2\left(\frac{h^2}{a^2} + \frac{k^2}{a^2} + \frac{l^2}{c^2}\right) = 1$$

$\therefore$
$$d^2\left(\frac{h^2 + k^2}{a^2} + \frac{l^2}{c^2}\right) = 1$$

Hence,
$$d_{hkl} = \left(\frac{h^2 + k^2}{a^2} + \frac{l^2}{c^2}\right)^{-\frac{1}{2}}$$

Crystal directions and planes :

In Fig. 1.25, the vector $\vec{r}$ passing through origin O and a lattice point is shown. The position vector $\vec{r}$ can be expressed in terms of fundamental translation vectors $\vec{a}$, $\vec{b}$ and $\vec{c}$ which form the crystal axes.

$$\vec{r} = n_1\,\vec{a} + n_2\,\vec{b} + n_3\,\vec{c}$$

where n_1, n_2, n_3 are integers.

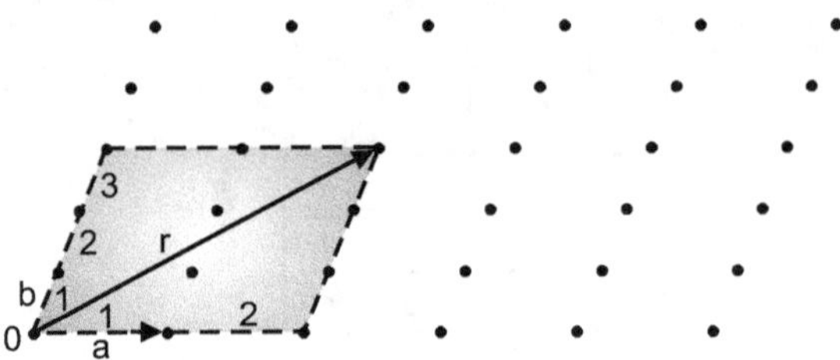

Fig. 1.25 : The Miller indices of the crystal direction denoted by vector r are [230]

- The c-axis is not shown in figure because $\vec{r}$ is lying in ab plane. The components of $\vec{r}$ along the three axes are $n_1 = 2$, $n_2 = 3$ and $n_3 = 0$. Then the crystal direction denoted by $\vec{r}$ is written as [230] in the Miller notation, with square brackets enclosing indices. If there is negative component along a crystal axis such as –3, it is written as $\bar{3}$ and read as bar 3.

1.10 SC, BCC and FCC Structures

Among the seven crystal systems, the cubic crystal is the simplest. The unit cell of a cubic structure is a cube. There are three types of unit cells, namely (i) simple cubic (SC), (ii) body centred cubic (BCC) and (iii) face centred cubic cell (FCC).

(i) Simple Cubic structure : In this type of crystal structure, the unit cell has one atom at each corner of the cube as shown in Fig. 1.26 (a).

This is the simplest of all structures, yet is rarely found to occur.

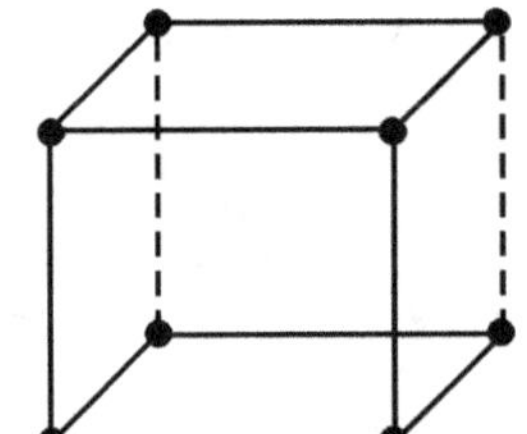

Fig. 1.26 (a) : Simple cubic unit cell

(ii) Body Centred Cubic structure :

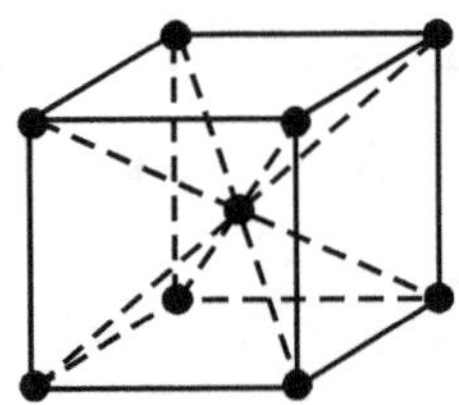

Fig. 1.26 (b) : BCC unit cell

In this type of crystal structure, the unit cell has one atom at each corner of the cube and one at body centre of the cube as shown in Fig. 1.26 (b).

Examples of this type of structure are α-iron, tungsten, molybdenum, chromium, alkali metals like sodium (Na), lithium (Li), potassium (K), rubidium (Rb), Caesium (Cs).

(iii) Face Centred Cubic structure :

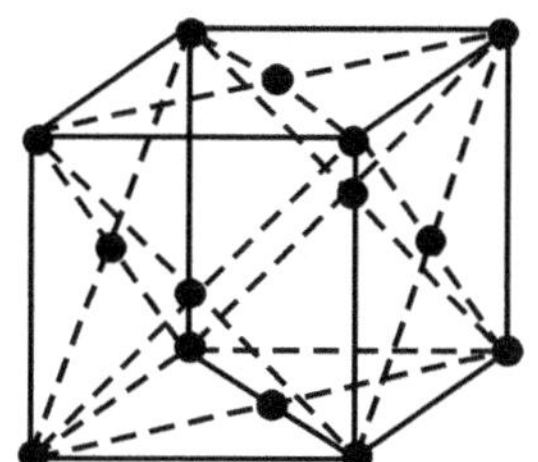

In this type of crystal structure, the unit cell has an atom at each corner of the cube and in addition has one atom at the centre of each face.

Examples of this type of structure are γ-iron, copper, silver, gold, aluminium, nickel, platinum etc.

Fig. 1.26 (c) : FCC unit cell

In this following, we shall obtain atomic radius, number of atoms per unit cell and co-ordination number for simple cubic, face centred cubic and body centred cubic systems.

1.10.1 Atomic Radius   (April 17)

The atomic radius is defined as half the distance between two nearest neighbouring atoms of same kind. It is denoted by r and is generally expressed in terms of cube edge 'a'.

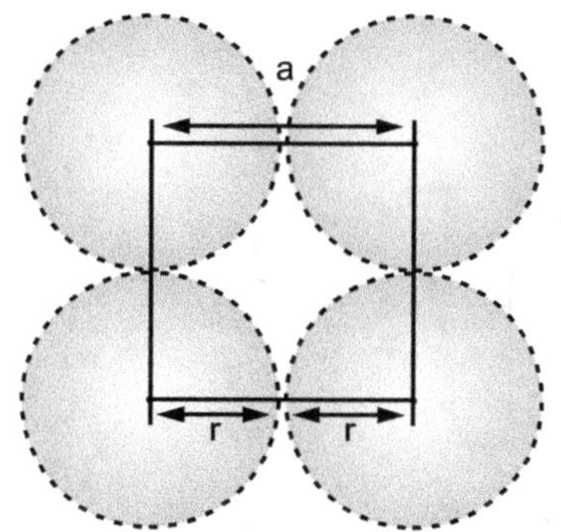

Fig. 1.27 (a) : Simple cubic system

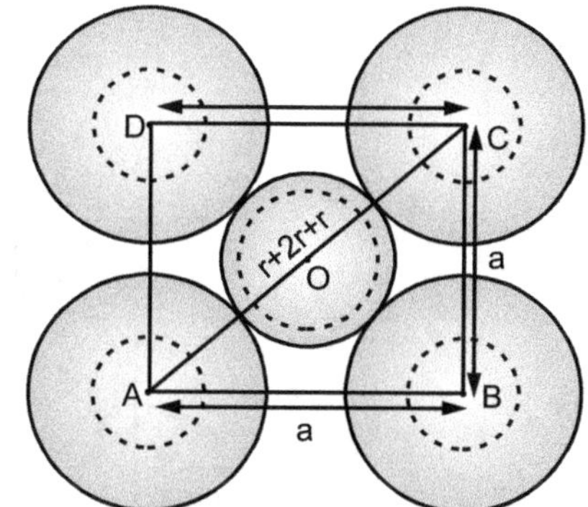

Fig. 1.27 (b) : Face centred cubic system

(1) Simple Cubic System (SC) :

For simple cubic system, atomic radius is r = a/2. In simple cubic system, edge length is 'a' which is the distance between the centres of two nearest neighbours as shown in Fig. 1.27 (a).

(2) Face Centred Cubic System (FCC) :   (April 17)

In face centred cubic system, the neighbouring atoms are at corners and an atom at the centre of adjacent faces as shown in Fig. 1.27 (b).

In Fig. 1.27 (b), A, B, C, D are the corner atoms and O is at the centre of adjacent face.

Therefore, atomic radius $r = \dfrac{OA}{2}$

In the right angled triangle $\triangle$ ABC,

$$AC^2 = AB^2 + BC^2$$

$$(r + 2r + r)^2 = a^2 + a^2$$

$$\therefore \quad (4r)^2 = 2a^2$$

$$r^2 = \frac{a^2}{8}$$

or

$$r = \frac{a}{2\sqrt{2}}$$

(3) Body Centred Cubic Crystal (BCC) :   (April 17)

In body centred cubic system the neighbouring atoms are at corners and there is one atom at the centre of cube as shown in Fig. 1.27 (c). From figure, the atomic radius is,

$$r = \frac{OA}{2}$$

Now, consider $\triangle$ ABC. $AC^2 = AB^2 + BC^2 = a^2 + a^2 = 2a^2$

$$\therefore \quad AC = \sqrt{2}\, a$$

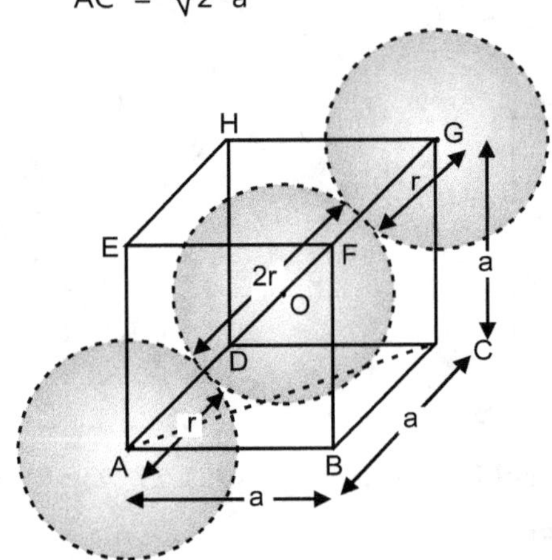

Fig. 1.27 (c) : Body centred cubic system

Also, from the geometry of figure,

$$AG = 4r$$

Consider right angled triangle $\triangle$ ACG.

$$AG^2 = AC^2 + GC^2$$

$$(4r)^2 = (\sqrt{2}a)^2 + a^2$$

$$16r^2 = 3a^2$$

$$\therefore \quad r = \frac{\sqrt{3}}{4}a$$

1.10.2 Number of Atoms in a Unit Cell   (April 17, Oct. 15)

The total number of atoms per unit cell is given by,

$$N = N_i + \frac{N_f}{2} + \frac{N_c}{8} \qquad \text{... (1.4)}$$

where,

$$N_i = \text{Number of body centred atoms}$$

$$N_f = \text{Number of face centred atoms}$$

$$N_c = \text{Number of corner atoms}$$

(1) Simple Cubic Crystal :

In simple cubic crystal system, there are only eight atoms present at corners. Therefore,

$$N_i = 0, N_f = 0, N_c = 8$$

Substituting these values in equation (1.4), we get

$$N = 0 + 0 + \frac{8}{8}$$

$$\therefore \qquad N = 1$$

Hence, for such crystal, there is one atom in each cell.

(2) Face Centred Cubic Crystal (FCC) :   (April 17, 16)

In such crystal there are eight atoms at eight corners and one atom at the centre of each six faces.

Therefore, $N_i = 0$, $N_f = 6$ and $N_c = 8$

Substituting these values in equation (1.3), we get

$$N = 0 + \frac{6}{2} + \frac{8}{8}$$

$$\therefore \qquad N = 4$$

In such crystal there are four atoms in each cell.

(3) Body Centred Cubic System (BCC) :   (April 17; Oct. 16, 15)

In such a crystal there are eight atoms at eight corners and one atom is present at the centre of the cube. Therefore, $N_i = 1$, $N_f = 0$, $N_c = 8$

Substituting these values in equation (1.4), we get

$$N = 1 + 0 + \frac{8}{8}$$

$$N = 2$$

Hence, each cell has two atoms.

1.10.3 Co-ordination Number   (Oct. 15)

The number of nearest neighbours that an atom has in a unit cell is called as co-ordination number. Greater is the coordination number, the more closely packed up will be the structure.

(1) Simple Cubic Structure :

Its co-ordination number is six. It is so because each corner atom is linked with seven other unit cells that can be imagined to be build around the unit cell containing the atom. In that case each corner atom has four neighbours in the same plane, one vertically above and one immediately below, giving a total number of six nearest neighbouring atoms.

(2) BCC Structure :

Its co-ordination number is eight. In this case the nearest neighbours of any corner atom are the body centred atoms. Since there are eight surrounding unit cells for any corner atom, their eight body centred atoms form the nearest neighbours for any corner atom.

(3) FCC Structure :

Its co-ordination number is twelve. In this case, nearest neighbours of any corner atom are the face centred atoms of the surrounding unit cells. Any corner atom has four such atoms in its own plane, four in a plane above it and four in a plane below it.

Table 1.2 indicates all above parameters for SC, BCC and FCC structures.

Table 1.2

	SC	BCC	FCC
Co-ordination number	6	8	12
Atomic radius (r)	$\dfrac{a}{2}$	$\dfrac{\sqrt{3} \cdot a}{4}$	$\dfrac{\sqrt{2} \cdot a}{4}$
Atoms per unit cell	1	2	4
Density of packing	$\dfrac{\pi}{6}$	$\dfrac{\sqrt{3} \cdot \pi}{8}$	$\dfrac{\sqrt{2} \cdot \pi}{6}$

1.11 Packing Fraction (PF)   (Oct. 17, 16; April 17, 16)

"Packing fraction is defined as the ratio of volume of atoms occupying the unit cell to the total volume of the unit cell relating to that structure".

$$\text{Packing Fraction (PF)} = \frac{\text{Number of atoms in unit cell} \times \text{Volume of each atom}}{\text{Volume of unit cell}}$$

The packing fraction gives us an idea about how closely atoms are packed together in a given crystal system. A low value of PF indicates loose packing of atoms and there is more unoccupied space. A higher value of PF indicates that the atoms are very closely packed in unit cell. The packing fraction for some cubic systems can be obtained as follows :

(1) Simple Cubic System : (April 17)

For simple cubic system, number of atoms per unit cell is one. The radius of atom is r and a is the edge length of the cube. Volume of unit cell=a^3 If atomic radius r is equal to a/2 then,

$$\text{Volume of one atom} = \frac{4}{3}\pi r^3 = \frac{4}{3}\pi \left(\frac{a}{2}\right)^3 = \frac{\pi a^3}{6}$$

Therefore,

$$PF = (1)\frac{\pi a^3/6}{a^3} = \frac{\pi}{6} = 0.52 \text{ or } 52\%$$

Only one polonium at a certain temperature exhibits this structure. SC structure is loosely packed structure.

(2) Body Centred Cubic (BCC) :

Elements like Li, Na, K and Cr exhibit this structure. For body centred cubic system the number of atoms per unit cell is two.

Atomic radius,

$$r = \frac{\sqrt{3}}{4}a$$

$$\text{Volume of one atom} = \frac{4}{3}\pi r^3 = \frac{4}{3}\pi \left(\frac{\sqrt{3}}{4}a\right)^3 = \frac{\pi a^3 \sqrt{3}}{16}$$

$$PF = (2) \times \frac{\pi a^3 \sqrt{3}}{16a^3} = \frac{\sqrt{3}\,\pi}{8} = 0.68 \text{ or } 68\%$$

BCC structure is a closely packed structure.

(3) Face Centred Cubic (FCC) :

For this system the number of atoms per unit cell is four.

$$\text{Atom radius, } r = \frac{a}{2\sqrt{2}}$$

$$\text{Volume of one atom} = \frac{4}{3}\pi r^3 = \frac{4}{3}\pi \left(\frac{a}{2\sqrt{2}}\right)^3$$

$$\text{Therefore, Volume of 4 atoms} = 4 \times \frac{4}{3} \times \left(\frac{a}{2\sqrt{2}}\right)^3 = \frac{\pi a^3}{3\sqrt{2}}$$

Hence,

$$\text{Packing Fraction PF} = \frac{\pi a^3}{\sqrt{2}\,a^3} = \frac{\pi}{3\sqrt{2}} = 0.74 \text{ or } 74\%$$

Copper, aluminium, lead and silver have this structure.

- From above three cases, it is found that in simple cubic system, 52% space is occupied by the atoms while empty space is 48%. Thus, in this case atoms are loosely packed. In BCC system, 58% space is occupied by the atoms and 32% is empty space. Therefore, in BCC system, atoms are more closely packed as compared to simple cubic system.

- In case of FCC system, 74% space is occupied by atoms and 26% is empty space. Hence, in this system atoms are more closely packed as compared to simple cubic system and BCC system.

Density of Crystal Material :

It is the ratio of mass of unit cell to the volume of unit cell.

Therefore, $$I = \frac{\text{Mass of unit cell}}{\text{Volume of unit cell}}$$

Let A be atomic weight of the element and N_o be Avogadro's number.

Then, Mass of one atom $= \dfrac{A}{N_o}$

If there are n number of atoms in unit cell, then mass of unit cell $= \dfrac{nA}{N_o}$, and volume of single cubic cell is a^3. Then, density $I = \dfrac{nA}{N_o \, a^3}$.

1.12 Some Crystal Structures

1. **Sodium Chloride Structure (NaCl) :**

The structure of sodium chloride is shown in Fig. 1.28 (a).

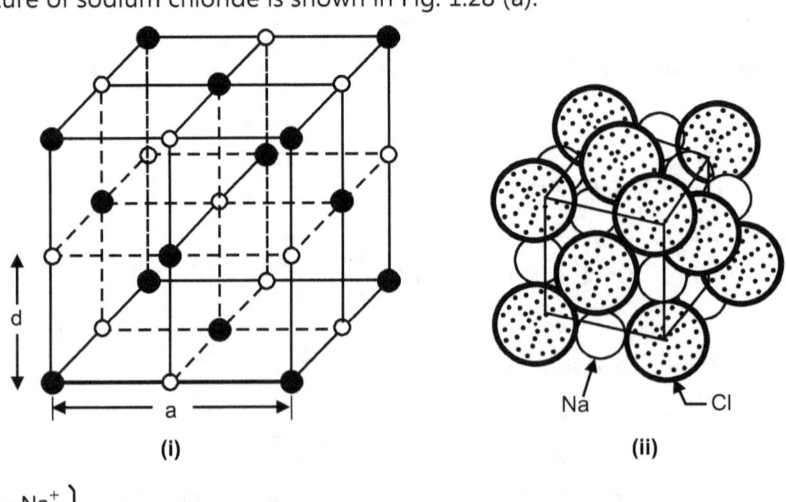

Fig. 1.28 : Sodium chloride structure

It is a face centred cubic lattice. The basis consists of one Na^+ ion and one Cl^- ion separated by one half the body diagonal of a unit cube. Hence, it may be considered as a unit cube. It may be considered as two face centred cubic, sub-lattices, one of Na^+ ion having its origin at the point (0, 0, 0) and other of Cl^- ion having its origin mid-way along a cube edge at a point $\left(\dfrac{a}{2}, 0, 0\right)$. There are four molecules of NaCl in each unit cube with ions in positions as given below.

$$Na^+ \qquad \dfrac{1}{2}\dfrac{1}{2}\dfrac{1}{2} \qquad 0\,0\,\dfrac{1}{2} \qquad 0\,\dfrac{1}{2}\,0 \qquad \dfrac{1}{2}\,0\,0$$

$$Cl^- \qquad 000 \qquad \dfrac{1}{2}\dfrac{1}{2}\,0 \qquad \dfrac{1}{2}\,0\,\dfrac{1}{2} \qquad 0\,\dfrac{1}{2}\dfrac{1}{2}$$

In this structure, each ion is surrounded by six nearest neighbours of opposite kind. The coordination number is six. There are twelve next nearest neighbours of some kind as the reference ion. The crystals having similar structure of NaCl are KCl, LiH, KBr, PbS, MgO, BaO etc.

2. Caesium Chloride Structure (CsCl) :

The caesium chloride structure (CsCl) is shown in Fig. 1.29. The space lattice is simple cubic with one molecule per unit cell. The basis has one Cs^+ ion at the position (0, 0, 0) and Cl^- ion at position $\left(\dfrac{1}{2}, \dfrac{1}{2}, \dfrac{1}{2}\right)$. In the unit cell of this type, each ion is at the centre of a cube of ions of the opposite kind, so the co-ordination number is eight. In Fig. 1.29, Cs^+ ions are at the eight corners while Cl^- ion is at the centre of a body of unit cell. The lattice points of CsCl are two interpenetrating simple cubic lattices, the corner of one sub lattice is the body centre of the other. So one sub-lattice is occupied by Cs^+ ion and other by Cl^- ion. RbCl and LiHg are some materials crystallizing in this structure.

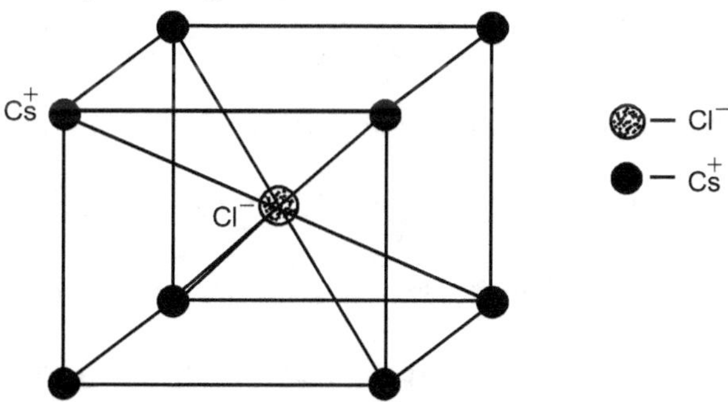

Fig. 1.29 : CsCl structure

3. Diamond Cubic Structure :

Diamond exhibits both cubic and hexagonal type structures. The diamond cubic structure is more common and so it is described here.

The space lattice of the diamond cubic structure may be regarded as a superposition of an FCC lattice and another FCC lattice obtained by translating the first lattice along the body diagonal by $\frac{1}{4}$ of the diagonal. The unit cell is shown in Fig. 1.30.

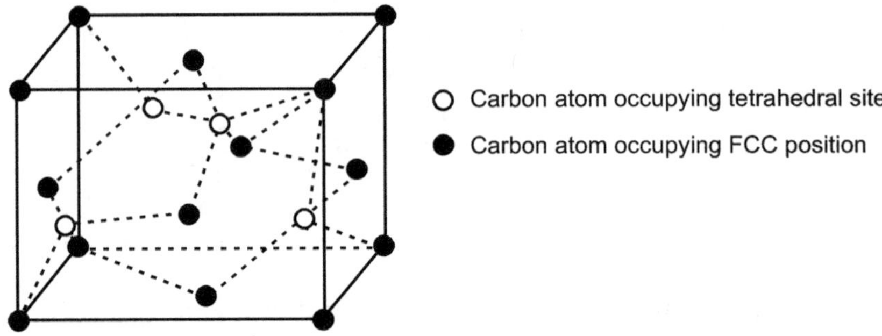

Fig. 1.30 : Diamond structure

In figure dark circles represent the FCC lattice and open circles represent the FCC lattice obtained by translating the first lattice along the body diagonal by $\frac{1}{4}^{\text{th}}$ of the diagonal.

A plane view of the positions of all the carbon atoms in the unit cell is shown in Fig. 1.31.

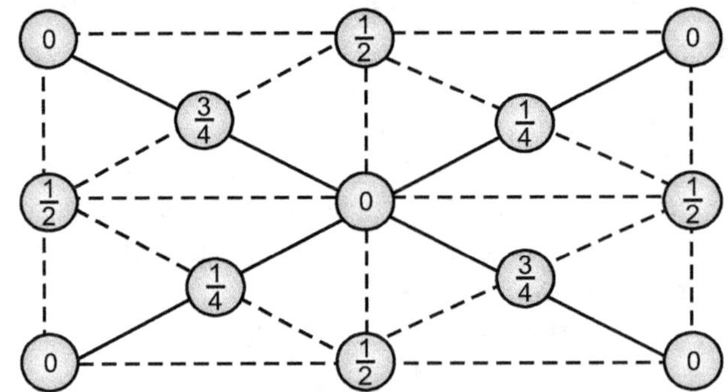

Fig. 1.31 : Plane view of atomic position in diamond unit cell

The points shown by 0 and $\frac{1}{2}$ are on the FCC lattice and at $\frac{1}{4}$ and $\frac{3}{4}$ are on a similar lattice displaced along the body diagonal by a quarter $\left(\frac{1}{4}\right)$ of the edge of the cube.

The materials having this type of structure are carbon, silicon and germanium. The packing fraction of this structure is 0.34 as compared to 0.74 of the FCC structure. Thus it is loosely packed structure.

4. Zinc Blende (ZnS) Structure :   (April 17)

The zinc blende (ZnS) structure is almost identical to diamond structure except that the two FCC lattices in it are occupied by different elements. The cubic zinc sulfide structure results when Zn atoms are placed on one FCC lattice and S atoms are on another FCC lattice. So the structure is a cube as shown in Fig. 1.32, where dark circles are of one type of atoms say sulphur (S) and open circles are of other type of atoms say zinc (Zn). The coordinates of Zn atoms are $000, 0\frac{1}{2}\frac{1}{2}, \frac{1}{2}0\frac{1}{2}, \frac{1}{2}\frac{1}{2}0$. The coordinates of S atoms are $\frac{1}{4}\frac{1}{4}\frac{1}{4}, \frac{1}{4}\frac{3}{4}\frac{1}{4}, \frac{3}{4}\frac{1}{4}\frac{3}{4}, \frac{3}{4}\frac{3}{4}\frac{1}{4}$. The lattice is FCC. There are four molecules of ZnS per conventional cell. About each atom there are four equally distant atoms of the opposite kind arranged at the corners of a regular tetrahedron.

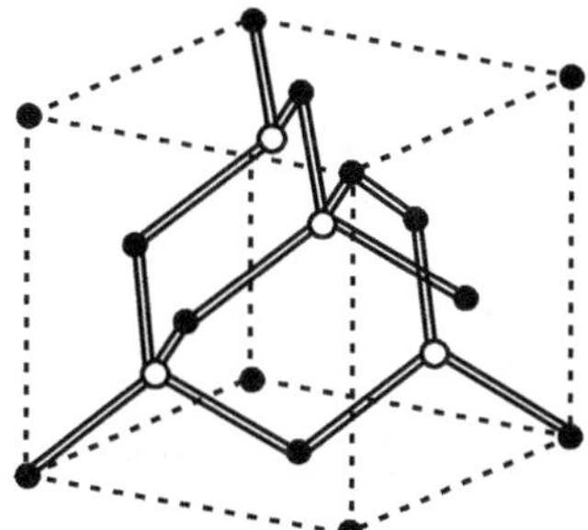

Fig. 1.32 : Crystal structure of cubic zinc sulfide

The compounds which has cubic zinc sulfide structure are CuCl, InSb and CdS.

5. **The hexagonal close packed (hcp) structure :**   **(April 17)**

The structure is built up stacking close-packed planes in a simple sequence.

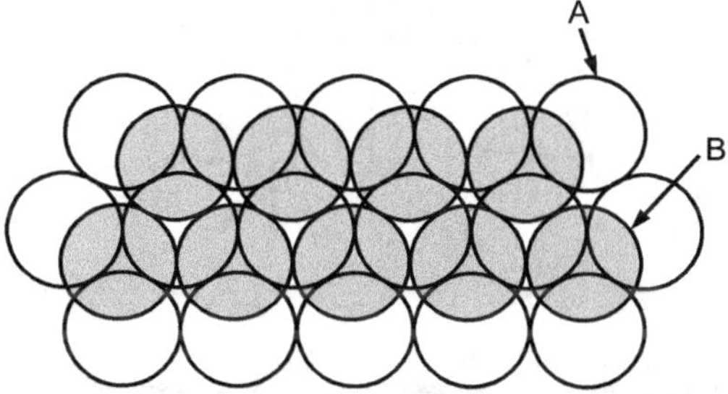

Fig. 1.33

Consider a layer of similar atoms (denoted by A in Fig. 1.33) with each atom surrounded by six atoms in one plane. Another layer B of atoms is placed on the top of the layer of atoms A in such a way that the atom of layer B occupy the alternate valleys formed by the atoms of layer A. Now a third layer of atoms can be placed on top of the B-layer in such a way that the atomic positions of third layer exactly coincide (or overlap) with atomic position of first layer denoted by A in Fig. 1.33. This type of stacking can be repeated successively resulting the following layered arrangement of atoms … ABABAB …. This type of stacking is called hcp stacking and the resulting structure is known as hexagonal close packed structure. The name suggests that the shape of conventional unit cell is hexagonal as shown in Fig. 1.34.

From Fig. 1.34, there are 12 atoms situated at the corners of unit cell, two atoms at the centre of the basal planes and three atoms completely situated inside the hexagonal unit cell of B layer. The effective number of atoms per unit cell is given by

= Contribution from corner atoms + Contribution from basal plane atoms

+ Contribution from the atoms inside the body

$$= 12 \times \frac{1}{6} + 2 \times \frac{1}{2} + 3$$

$$= 2 + 1 + 3 = 6$$

Thus in hexagonal closed-packed structure, there are 6 atoms per unit cell.

This structure is called close-packed because it possesses maximum density of packing of hard spheres. For an ideal hcp structure, the packing fraction is 0.74. Each atom in this structure has 12 nearest neighbours, so it's co-ordination number is 12.

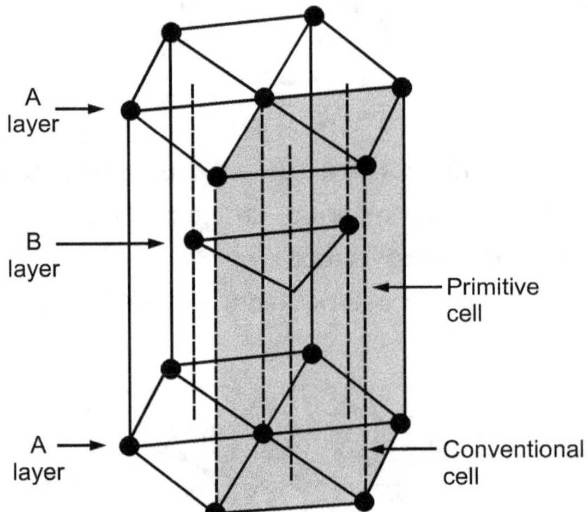

Fig. 1.34 : Hexagonal closed packed structure

Magnesium, zinc and cadmium crystallizes in this structure.

1.13 Concept of Reciprocal Lattice

- The crystal lattice is conveniently represented by reciprocal lattice. The reciprocal lattice concept is very useful in the study of X-ray diffraction pattern produced by a crystal. The reciprocal lattice of a real crystal is constructed as follows :
    1. Select a point in direct lattice as origin.
    2. From this common origin draw normals to each and every set of parallel planes in the direct lattice.
    3. Fix the length of each normal equal to the reciprocal of the interplaner spacing of that set of parallel planes (h k l) it represents.
    4. Place a point at the end of the normal.

The array of point thus generated, makes the reciprocal lattice of the crystal under consideration.

Now, consider a two-dimensional square lattice as shown in Fig. 1.35 (a). The origin O is shown in Fig. 1.35 (b).

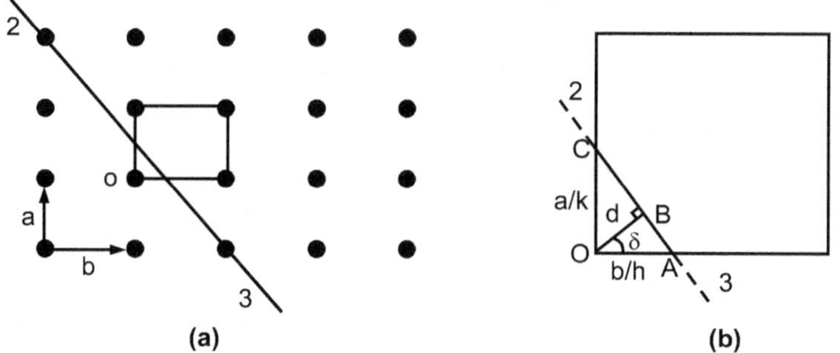

(a)   (b)

**Fig. 1.35 : (a) A square lattice showing the unit cell and a (h, k) plane,
(b) the imagined unit cell**

Consider a plane represented by the line (32) in Fig. 1.35 (a). By the property of similarity of triangles [Refer Fig. 1.35 (b)], we get

$$\frac{OB}{OA} = \frac{OC}{AC}$$

$$\frac{d}{b/h} = \frac{a/k}{\sqrt{\dfrac{a^2}{k^2} + \dfrac{b^2}{h^2}}}$$

$$\therefore \qquad \frac{1}{d} = \frac{\sqrt{h^2\,a^2 + k^2\,b^2}}{ab} \qquad\qquad \text{... (1.5)}$$

If $\sigma = \dfrac{1}{d}$, then equation (1.5) becomes,

$$\sigma = \frac{\sqrt{h^2\,a^2 + k^2\,b^2}}{ab} \qquad\qquad \text{... (1.6)}$$

The direction of σ is,

$$\tan \delta = \frac{b/h}{a/k}$$

The line LM with length equal to $\dfrac{1}{d}$ or $\sigma = \dfrac{\sqrt{h^2a^2 + k^2b^2}}{ab}$ along the direction coinciding with normal is shown in Fig. 1.36. The terminal points of this line are the reciprocal lattice points.

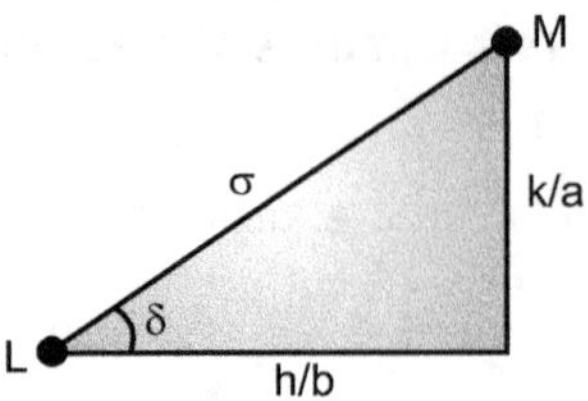

Fig. 1.36 : The reciprocal lattice point

The planes of a square lattice are represented by a square reciprocal lattice and so on. The diffraction of X-rays by a real crystal lattice will give an array of points identical with its reciprocal lattice.

Reciprocal Lattice

Let σ_{hkl} represent the reciprocal lattice vector. **It is defined as a vector having magnitude equal to the reciprocal of the interplaner spacing d_{dkl} and direction coinciding along normal to the (hk*l*) planes.**

Hence,
$$|\vec{\sigma}_{hkl}| = \frac{1}{d_{hkl}} \qquad \qquad ... (1.7)$$

where, d_{hkl} is the interplaner distance.

The magnitude of σ_{hkl} is $\dfrac{1}{d_{hkl}}$ and its direction is parallel to the normal to the hk*l* planes.

Consider a primitive unit cell of the crystal lattice having crystallographic axes a, b and c as shown in Fig. 1.37.

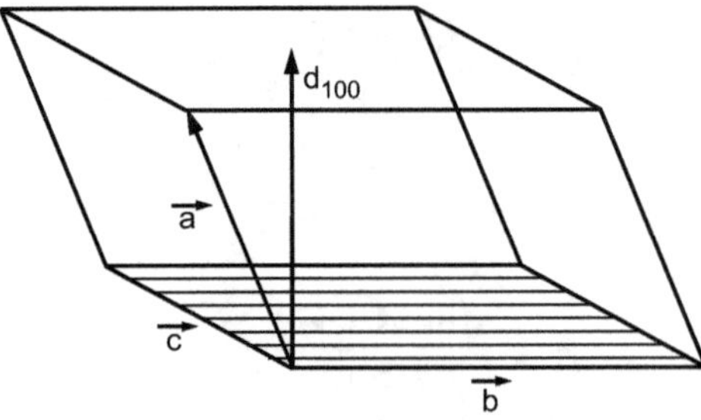

Fig. 1.37 : Primitive unit cell

We have to find the relation between normal to the plane and a, b, c. The volume of this primitive cell V is equal to the product of area of base having sides b and c and the height of cell d_{100}.

$$\therefore \qquad V = \text{Area} \times d_{100}$$

$$\therefore \qquad \frac{1}{d_{100}} = \frac{\text{Area}}{V} \qquad \qquad ... (1.8)$$

If $\hat{n}$ is a unit vector in the direction of normal to the plane, then equation (1.6) can be expressed as,

$$\vec{\sigma}_{hkl} = \frac{1}{d_{hkl}} \, \hat{n} \qquad \qquad \text{... (1.9)}$$

The area of the base of primitive cell is,

$$\text{Area} = |\vec{b} \times \vec{c}| \qquad \qquad \text{... (1.10)}$$

Using equation (1.10) in equation (1.8), we get,

$$\frac{1}{d_{100}} = \frac{|\vec{b} \times \vec{c}|}{V}$$

Hence,

$$\vec{\sigma}_{100} = \frac{|\vec{b} \times \vec{c}|}{V} \, \hat{n} \qquad \qquad \because \hat{n} = \frac{\vec{b} \times \vec{c}}{|\vec{b} \times \vec{c}|}$$

$$= \frac{\vec{b} \times \vec{c}}{V}$$

The scalar triple product of three vectors $\vec{a}$, $\vec{b}$ and $\vec{c}$ gives the volume of primitive cell.

$$\therefore \qquad V = \vec{a} \cdot (\vec{b} \times \vec{c})$$

$$\therefore \qquad \vec{\sigma}_{100} = \frac{\vec{b} \times \vec{c}}{\vec{a} \cdot (\vec{b} \times \vec{c})} \qquad \qquad \text{... (1.11)}$$

Similarly, we can obtain

$$\vec{\sigma}_{010} = \frac{\vec{c} \times \vec{a}}{\vec{a} \cdot (\vec{b} \times \vec{c})} \qquad \qquad \text{... (1.12)}$$

$$\vec{\sigma}_{001} = \frac{\vec{a} \times \vec{b}}{\vec{a} \cdot (\vec{b} \times \vec{c})} \qquad \qquad \text{... (1.13)}$$

The vectors $\vec{\sigma}_{100}$, $\vec{\sigma}_{010}$ and $\vec{\sigma}_{001}$ are considered as the reciprocal translation vectors. These vectors can be represented by,

$$\vec{A} = \vec{\sigma}_{100} = \frac{\vec{b} \times \vec{c}}{\vec{a} \cdot (\vec{b} \times \vec{c})}$$

$$\vec{B} = \sigma_{010} = \frac{\vec{c} \times \vec{a}}{\vec{a} \cdot (\vec{b} \times \vec{c})}$$

and

$$\vec{C} = \sigma_{001} = \frac{\vec{a} \times \vec{b}}{\vec{a} \cdot (\vec{b} \times \vec{c})} \qquad \ldots (1.14)$$

The relation between reciprocal translation vectors $\vec{A}$, $\vec{B}$, $\vec{C}$ and translation vectors $\vec{a}$, $\vec{b}$, $\vec{c}$ may be written as,

$\vec{A}$ is normal to $\vec{b}$ and $\vec{c}$.

$\vec{B}$ is normal to $\vec{c}$ and $\vec{a}$.

$\vec{C}$ is normal to $\vec{a}$ and $\vec{b}$.

$$\therefore \qquad \vec{A} \cdot \vec{b} = 0 \qquad \vec{A} \cdot \vec{c} = 0$$

$$\vec{B} \cdot \vec{c} = 0 \qquad \vec{B} \cdot \vec{a} = 0$$

$$\vec{C} \cdot \vec{a} = 0 \qquad \vec{C} \cdot \vec{b} = 0$$

With the help of equation (1.14), we get,

$$\vec{A} \cdot \vec{a} = \left[\frac{\vec{b} \times \vec{c}}{\vec{a} \cdot (\vec{b} \times \vec{c})} \right] \cdot \vec{a} = \frac{\vec{a} \cdot (\vec{b} \times \vec{c})}{\vec{a} \cdot (\vec{b} \times \vec{c})} = 1 \qquad \ldots (1.15)$$

Similarly,

$$\vec{B} \cdot \vec{b} = \left[\frac{\vec{c} \times \vec{a}}{\vec{a} \cdot (\vec{b} \times \vec{c})} \right] \cdot \vec{b} = \frac{\vec{b} \cdot (\vec{c} \times \vec{a})}{\vec{a} \cdot (\vec{b} \times \vec{c})}$$

But from vector analysis the scalar triple product

$$\vec{a} \cdot (\vec{b} \times \vec{c}) = \vec{b} \cdot (\vec{c} \times \vec{a}) = \vec{c} \cdot (\vec{a} \times \vec{b}) \qquad \ldots (1.16)$$

$$\therefore \qquad \vec{b} \cdot (\vec{c} \times \vec{a}) = \vec{a} \cdot (\vec{b} \times \vec{c})$$

$$\therefore \qquad \vec{B} \cdot \vec{b} = \frac{\vec{a} \cdot (\vec{b} \times \vec{c})}{\vec{a} \cdot (\vec{b} \times \vec{c})} = 1 \qquad \ldots (1.17)$$

and
$$\vec{C} \cdot \vec{c} = \left[\frac{\vec{a} \times \vec{b}}{\vec{a} \cdot (\vec{b} \times \vec{c})} \right] \cdot \vec{c} = \frac{\vec{c} \cdot (\vec{a} \times \vec{b})}{\vec{a} \cdot (\vec{b} \times \vec{c})}$$

But from equation (1.17),

$$\vec{c} \cdot (\vec{a} \times \vec{b}) = \vec{a} \cdot (\vec{b} \times \vec{c})$$

$$\therefore \qquad \vec{C} \cdot \vec{c} = \frac{\vec{a} \cdot (\vec{b} \times \vec{c})}{\vec{a} \cdot (\vec{b} \times \vec{c})} = 1 \qquad\qquad \text{... (1.18)}$$

With the help of reciprocal lattice vectors $\vec{A}$, $\vec{B}$ and $\vec{C}$, we can construct a lattice. Then in this case the successive points in the $\vec{A}$ direction show successive multiples h of the spacing of (100); in the $\vec{B}$ direction successive multiples of k of the spacing of (010); and in the $\vec{C}$ direction, successive multiples l of the spacing of (001).

Therefore, in order to obtain a reciprocal lattice point (hk$l$), one has to go h units along $\vec{A}$, k units along vector $\vec{B}$ and l units along $\vec{C}$. Hence, the reciprocal lattice vector $\vec{\sigma}_{hk}$ will be,

$$\vec{\sigma}_{hkl} = h\vec{A} + k\vec{B} + l\vec{C}$$

1.14 Properties of the Reciprocal Lattice   (Oct. 16)

(1) The reciprocal of the reciprocal lattice is the direct lattice.

In order to prove this property, consider reciprocal lattice vector

$$\vec{A} = \frac{\vec{b} \times \vec{c}}{\vec{a} \cdot (\vec{b} \times \vec{c})} \qquad\qquad \text{... (1.19)}$$

Let $\vec{A} = \vec{a^*}$, $\vec{B} = \vec{b^*}$, $\vec{C} = \vec{c^*}$.

Therefore, equation (1.19) becomes,

$$\vec{a^*} = \frac{\vec{b} \times \vec{c}}{\vec{a} \cdot (\vec{b} \times \vec{c})} \qquad\qquad \text{... (1.20)}$$

Taking its reciprocal,

$$(\vec{a^*})^* = \frac{\vec{b^*} \times \vec{c^*}}{\vec{a^*} \cdot (\vec{b^*} \times \vec{c^*})} \qquad\qquad \text{... (1.21)}$$

Multiply RHS of equation (1.21) by $\vec{a} \cdot \vec{a}^*$

Since, $(\vec{a} \cdot \vec{a}^* = 1)$

Therefore, we get,

$$(\vec{a}^*)^* = \vec{a} \cdot \vec{a}^* \left[\frac{\vec{b}^* \times \vec{c}^*}{\vec{a}^* \cdot (\vec{b}^* \times \vec{c}^*)} \right]$$

$$= \left[\frac{\vec{a}^* \cdot (\vec{b}^* \times \vec{c})}{\vec{a}^* \cdot (\vec{b}^* \times \vec{c}^*)} \right] \vec{a}$$

$$= \vec{a} \qquad \qquad \qquad \dots (1.22)$$

Therefore, from equations (1.20) and (1.21), we get

$$\vec{a} = \frac{\vec{b}^* \times \vec{c}^*}{\vec{a}^* \cdot (\vec{b}^* \times \vec{c}^*)}$$

or

$$\vec{a} = \frac{\vec{B} \times \vec{C}}{\vec{A} \cdot (\vec{B} \times \vec{C})}$$

Similarly,

$$\vec{b} = \frac{\vec{C} \times \vec{A}}{\vec{A} \cdot (\vec{B} \times \vec{C})}$$

$$\vec{c} = \frac{\vec{A} \times \vec{B}}{\vec{A} \cdot (\vec{B} \times \vec{C})}$$

(2) The volume of unit cell of the reciprocal lattice is inversely proportional to the volume of a unit cell of direct lattice.   **(Oct. 15)**

We know that,

$$\text{Volume of direct lattice} = \vec{a} \cdot (\vec{b} \times \vec{c})$$

and

$$\text{Volume of reciprocal lattice} = \vec{A} \cdot (\vec{B} \times \vec{C})$$

But we know,

$$\vec{A} = \frac{\vec{b} \times \vec{c}}{\vec{a} \cdot (\vec{b} \times \vec{c})}$$

$$\vec{B} = \frac{\vec{c} \times \vec{a}}{\vec{a} \cdot (\vec{b} \times \vec{c})}$$

and

$$\vec{C} = \frac{\vec{a} \times \vec{b}}{\vec{a} \cdot (\vec{b} \times \vec{c})}$$

$$\therefore \quad \vec{A} \cdot (\vec{B} \times \vec{C}) = \left[\frac{\vec{b} \times \vec{c}}{\vec{a} \cdot (\vec{b} \times \vec{c})}\right] \cdot \left[\frac{\vec{c} \times \vec{a}}{\vec{a} \cdot (\vec{b} \times \vec{c})}\right] \times \left[\frac{\vec{a} \times \vec{b}}{\vec{a} \cdot (\vec{b} \times \vec{c})}\right]$$

$$= \frac{1}{(\vec{a} \cdot \vec{b} \times \vec{c})^3} \left[(\vec{b} \times \vec{c}) \cdot (\vec{c} \times \vec{a}) \times (\vec{a} \times \vec{b})\right] \qquad \text{... (1.23)}$$

But from vector analysis,

$$(\vec{P} \times \vec{Q}) \times (\vec{R} \times \vec{S}) = \vec{Q}\,(\vec{P} \cdot \vec{R} \times \vec{S}) - \vec{P}\,(\vec{Q} \cdot \vec{R} \times \vec{S})$$

$$\therefore \quad (\vec{c} \times \vec{a}) \times (\vec{a} \times \vec{b}) = \vec{a}\,(\vec{c} \cdot \vec{a} \times \vec{b}) - \vec{c}\,(\vec{a} \cdot \vec{a} \times \vec{b})$$

Substituting this value in equation (1.23), we get,

$$\vec{A} \cdot (\vec{B} \times \vec{C}) = \frac{1}{(\vec{a} \cdot \vec{b} \times \vec{c})^3}\,(\vec{b} \times \vec{c}) \cdot [\vec{a}\,(\vec{c} \cdot \vec{a} \times \vec{b}) - \vec{c}\,(\vec{a} \cdot \vec{a} \times \vec{b})]$$

$$\text{... (1.24)}$$

But

$$\vec{c} \cdot (\vec{a} \times \vec{b}) = \vec{a} \cdot (\vec{b} \times \vec{c})$$

and

$$\vec{a} \cdot \vec{a} \times \vec{b} = 0$$

$$\therefore \quad \vec{A} \cdot \vec{B} \times \vec{C} = \frac{1}{(\vec{a} \cdot \vec{b} \times \vec{c})^3}\,(\vec{b} \times \vec{c}) \cdot [\vec{a}\,(\vec{a} \cdot \vec{b} \times \vec{c})]$$

$$= \frac{(\vec{a} \cdot \vec{b} \times \vec{c})^2}{[\vec{a} \cdot (\vec{b} \times \vec{c})]^3}$$

$$\therefore \quad \vec{A} \cdot (\vec{B} \times \vec{C}) = \frac{1}{\vec{a} \cdot (\vec{b} \times \vec{c})} \qquad \text{... (1.25)}$$

$$\therefore \quad \vec{a} \cdot (\vec{b} \times \vec{c}) = \frac{1}{\vec{A} \cdot (\vec{B} \times \vec{C})}$$

$$\frac{\text{Volume of}}{\text{direct lattice}} = \frac{1}{\text{Volume of reciprocal lattice}}$$

(3) Every reciprocal lattice vector is perpendicular to a direct lattice plane.

In order to prove that $\vec{\sigma}_{hkl}$ is normal to crystal plane (hkl), we have to show that the scalar product of $\vec{\sigma}_{hkl}$ and vector lying in the plane (hkl) is zero.

Consider the (hkl) plane as shown in Fig. 1.38. In this figure, the intercept on X-axis is,

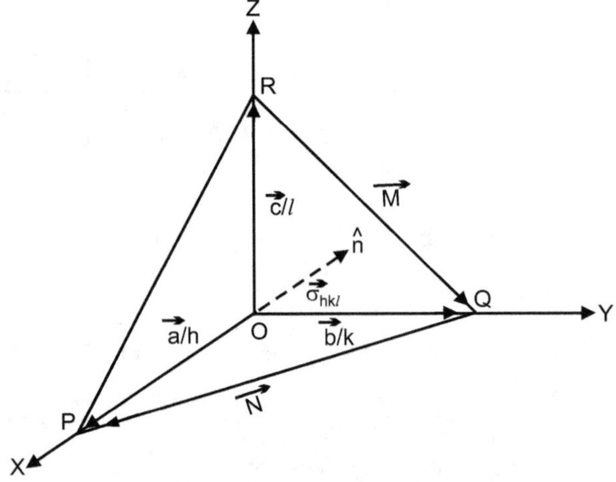

Fig. 1.38 : Direct lattice plane PQR

$$\vec{OP} = \frac{\vec{a}}{h}$$

$$\text{Intercept on Y-axis is } \vec{OQ} = \frac{\vec{b}}{k}$$

$$\text{and} \quad \text{Intercept on Y-axis is } \vec{OR} = \frac{\vec{c}}{l}$$

$$\vec{OQ} + \vec{QP} = \vec{OP}$$

From the geometry of Fig. 1.38,

$$\therefore \quad \frac{\vec{b}}{k} + \vec{N} = \frac{\vec{a}}{h}$$

$$\text{or} \quad \vec{N} = \frac{\vec{a}}{h} - \frac{\vec{b}}{k}$$

The scalar product of $\vec{N}$ and $\vec{\sigma}_{hkl}$ is given by,

$$\vec{N} \cdot \vec{\sigma}_{hkl} = \left(\frac{\vec{a}}{h} - \frac{\vec{b}}{k} \right) \cdot \vec{\sigma}_{hkl}$$

But
$$\vec{\sigma}_{hkl} = h\,\vec{A} + k\,\vec{B} + l\,\vec{C}$$

$\therefore$
$$\vec{N} \cdot \vec{\sigma}_{hkl} = \left(\frac{\vec{a}}{h} - \frac{\vec{b}}{k}\right) \cdot (h\,\vec{A} + k\,\vec{B} + l\,\vec{C})$$

$$= \frac{\vec{a}}{h} \cdot (h\,\vec{A} + k\,\vec{B} + l\,\vec{C})$$

$$-\frac{\vec{b}}{k} \cdot (h\,\vec{A} + k\,\vec{B} + l\,\vec{C})$$

But
$$\vec{a} \cdot \vec{A} = 1$$

$$= \left(\frac{h}{h} + 0 + 0\right) - \left(0 + \frac{k}{k} + 0\right)$$

$$= 1 - 1 \qquad\qquad\qquad \ldots (1.26)$$

$$= 0$$

Also, from Fig. 1.38,

$$\vec{OR} + \vec{RQ} = \vec{OQ}$$

$\therefore$
$$\frac{\vec{c}}{l} + \vec{M} = \frac{\vec{b}}{k}$$

or
$$\vec{M} = \frac{\vec{b}}{k} - \frac{\vec{c}}{l}$$

$$\vec{M} \cdot \vec{\sigma}_{hkl} = \left(\frac{\vec{b}}{k} - \frac{\vec{c}}{l}\right) \cdot (h\,\vec{A} + k\,\vec{B} + l\,\vec{C})$$

$$= \left(0 + \frac{k}{k} + 0\right) - \left(0 + 0 + \frac{l}{l}\right)$$

$$= 1 - 1$$

$$\vec{M} \cdot \vec{\sigma}_{hkl} = 0 \qquad\qquad\qquad \ldots (1.27)$$

From equations (1.26) and (1.27), we conclude that $\vec{\sigma}_{hkl}$ is normal to $\vec{M}$ and $\vec{N}$ vectors.

Hence, $\vec{\sigma}_{hkl}$ is normal to the plane containing $\vec{M}$ and $\vec{N}$ which is direct lattice plane (hkl).

Solved Examples

Example 1.1 : *A FCC crystal has an atomic radius of 1.246 A°. What are d_{200}, d_{220} and d_{111} spacings ?*

Solution : For FCC crystal the interatomic distance

$$a = \frac{4r}{\sqrt{2}}$$

$$= 2\sqrt{2}\, r$$

Given :   Atomic radius $r = 1.246$ A°

$\therefore$ $a = 2\sqrt{2} \times 1.246$

        $a = 3.524$ A°

For a crystal, $d_{hkl} = \dfrac{a}{\sqrt{h^2 + k^2 + l^2}}$

(i) $d_{200} = \dfrac{3.524}{\sqrt{2^2 + 0 + 0}} = \dfrac{3.524}{2}$

$\therefore$ $d_{200} = \mathbf{1.762\ A°}$ ... **Ans.**

(ii) $d_{220} = \dfrac{3.524}{\sqrt{4 + 4 + 0}} = \mathbf{1.245\ A°}$ ... **Ans.**

(iii) $d_{111} = \dfrac{3.524}{\sqrt{1 + 1 + 1}}$

$$= \mathbf{2.034\ A°}$$ ... **Ans.**

Example 1.2 : *Find out the number of atoms per square millimeter on a plane (100) of lead whose interatomic distance is 3.499 A°. Lead has face-centred cubic structure.*   **(Oct. 16)**

Solution :

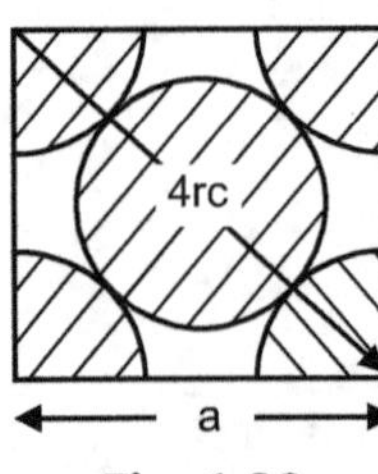

Fig. 1.39

Fig. 1.39 represents (100) plane. The shaded portion shows the portion of the atom.

For FCC structure,

$$r = \frac{\sqrt{2}\,a}{4}$$

$\therefore$ $a = \dfrac{4r}{\sqrt{2}} = 2\sqrt{2}\, r$

Interatomic distance $= 3.499$

$\therefore$ $2r = 3.499$

Now,
$$a = \sqrt{2}\,(3.499) = 4.95 \; A^{\circ}$$

$$\text{Area of plane} = a^2$$

where, a is side of given plane

$$= (4.95 \times 10^{-7})^2 \; mm^2$$

$$\text{Number of atoms in the plane} = 2$$

$$\text{Atoms/mm}^2 = \frac{2 \times 1}{(4.95 \times 10^{-7})^2} = 8.2 \times 10^{12} \; \textbf{atoms/mm}^2 \qquad \textbf{... Ans.}$$

Example 1.3 : *Calculate the number of atoms per unit cell of metal having a lattice parameter of 2.9 A° and density 7.87 gram/cc. Atomic weight of the metal is 55.85 and Avogadro constant is 6.023 $\times 10^{23}$.*

Solution :   Density of crystal $\rho = \dfrac{nM}{a^3 \, N_A}$

Given : $a = 2.9 \; A^{\circ} = 2.9 \times 10^{-8}$ cm, $M = 55.85$, $N_A = 6.023 \times 10^{23}$, $\rho = 7.87$ gm/cc.

$$\therefore \qquad n = \frac{\rho \, a^3 \, N_A}{M}$$

$$= \frac{7.87 \times (2.9 \times 10^{-8})^3 \times 6.023 \times 10^{23}}{55.85}$$

$$n = \textbf{2 atoms} \qquad\qquad \textbf{... Ans.}$$

Hence, unit cell may be body centred cubic.

Example 1.4 : *Sketch (112) plane in simple cubic cell.*

Solution : A plane whose Miller indices are (112) has the following intercepts on the three axes.

$$\frac{1}{1}, \frac{1}{1}, \frac{1}{2}$$

This plane is shown in Fig. 1.40.

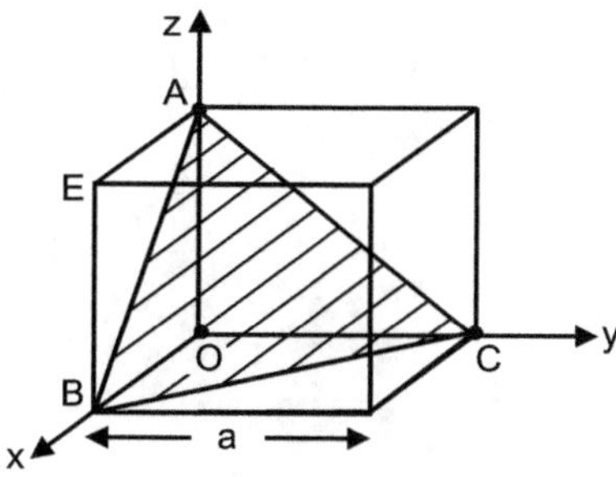

Here,

$$OB = OC = a$$

$$\text{and} \qquad OA = \frac{a}{2}$$

Fig. 1.40

Example 1.5 : *Sketch (110), (010), (001), (200), ($\bar{1}$00) and (112) planes in simple cubic cell.*

Solution : (i) A plane whose Miller indices are (100) has intercepts $\dfrac{1}{1}, \dfrac{1}{0}, \dfrac{1}{0}$

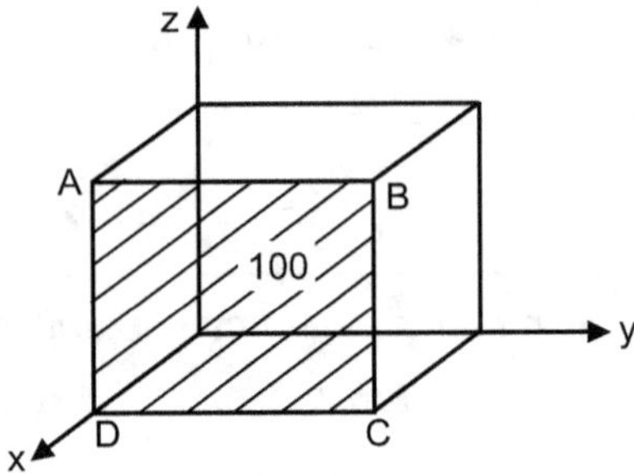

Fig. 1.41

(ii)   A plane whose Miller indices are (101) has intercepts

$$\frac{1}{1}, \frac{1}{0}, \frac{1}{1}$$

i.e. $1, \infty, 1$

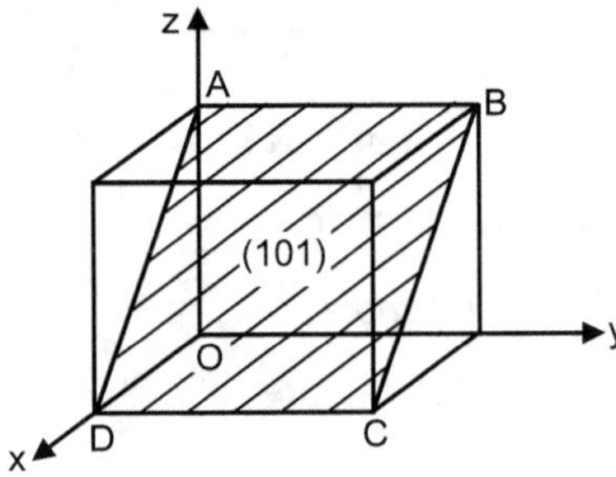

Fig. 1.42

(iii) A plane whose Miller indices are (110) has intercepts

$$\frac{1}{1}, \frac{1}{1}, \frac{1}{0}$$

i.e. $1, 1, \infty$

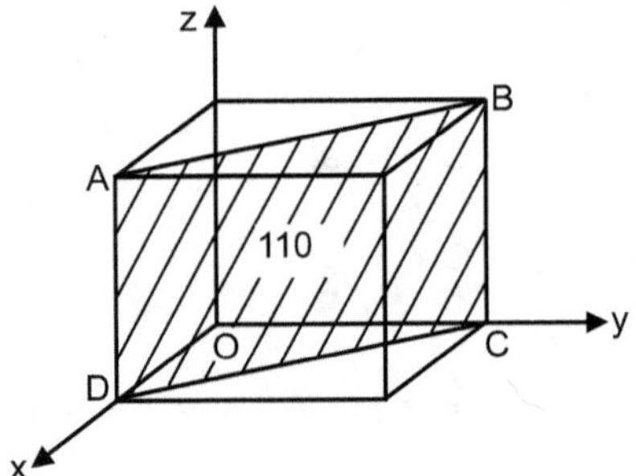

Fig. 1.43

(iv) A plane whose Miller indices are (010) has intercepts

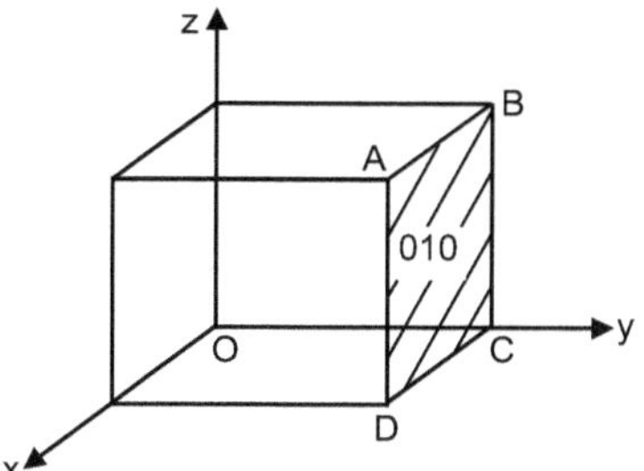

Fig. 1.44

$$\frac{1}{0}, \frac{1}{1}, \frac{1}{0}$$

i.e. $\infty, 1, \infty$

(v) A plane whose Miller indices are (001) has intercepts

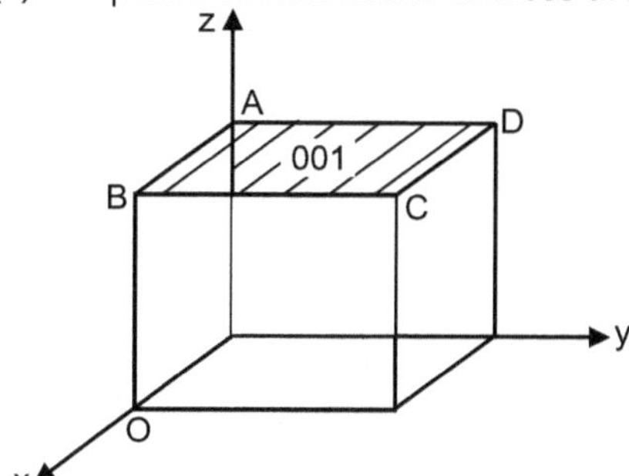

Fig. 1.45

$$\frac{1}{0}, \frac{1}{0}, \frac{1}{1}$$

i.e. $\infty, \infty, 1$

(vi) A plane whose Miller indices are (2, 0, 0) has intercepts

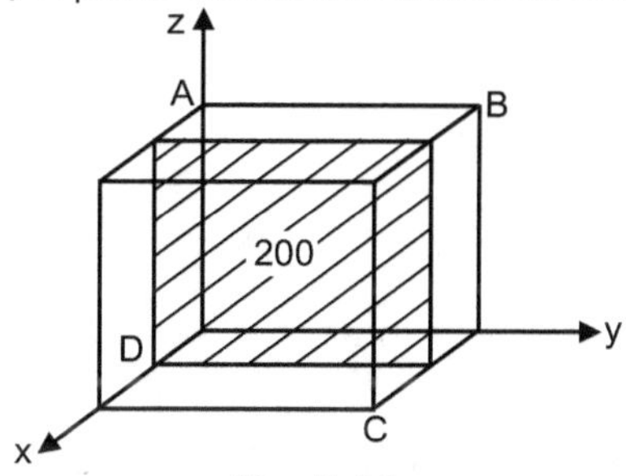

Fig. 1.46

$$\frac{1}{2}, \frac{1}{0}, \frac{1}{0}$$

i.e. $\frac{1}{2}, \infty, \infty$

(vii) A plane whose Miller indices are ($\bar{1}$, 0, 0) has intercepts

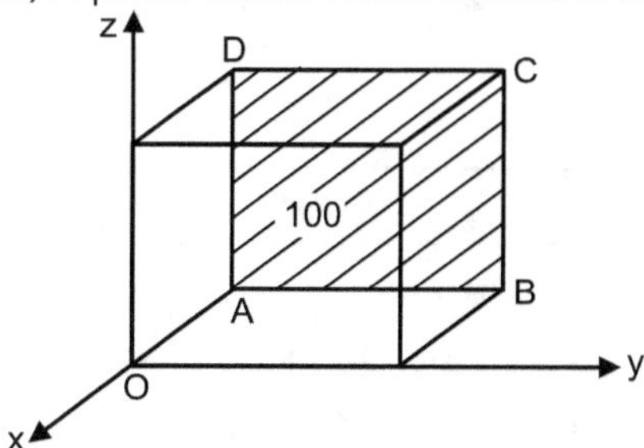

Fig. 1.47

$$\frac{-1}{1}, \frac{1}{0}, \frac{1}{0}$$

i.e. $-1, \infty, \infty$

Example 1.6 : *A crystal plane intercepts the three crystallographic axes at 3/2, 2 and 1. What are the Miller indices of the plane ?*

Solution : Given : Intercepts are

$$\frac{3}{2}, 2, 1$$

Take its reciprocal $\dfrac{2}{3}, \dfrac{1}{2}, 1$

Clear fractions by multiplying 6 are

$$4, 3, 6$$

Hence, Miller indices are (4, 3, 6)

Example 1.7 : *Determine the Miller indices for the plane in Fig. 1.48.*

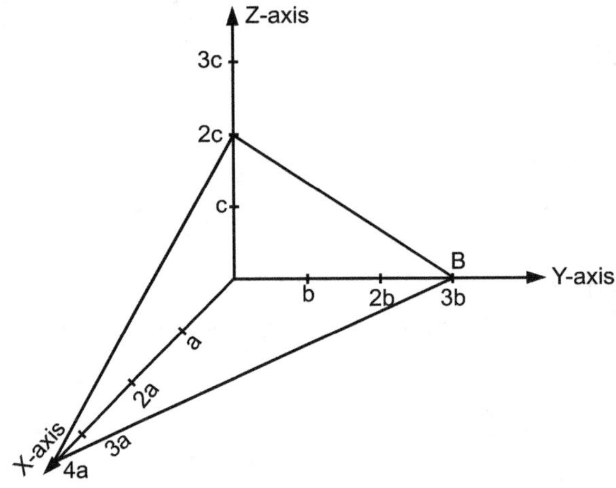

Fig. 1.48

Solution : The intersections with the three axes are at 4a, 3b and 2c. Then the inverse intercepts in lattice vector units are 1/4, 1/3, 1/2

To get integer numbers we have to calculate the lowest common denominator of this fraction, which is 12. Multiplying each fraction with 12 gives the three Miller Indices : (346).

Example 1.8 : *Calculate the Miller indices of crystal planes which cut through the crystal axes at (i) (2a, 3b, c), (ii) (6a, 3b, 3c) and (iii) (2a, − 3b, − 3c).*   **(April 16)**

Solution : Let us prepare the table as follows :

(i)

a	b	c	
2	3	1	intercepts
$\dfrac{1}{2}$	$\dfrac{1}{3}$	1	reciprocals
3	2	6	clear fractions

Hence, the Miller indices are (326).

(ii)

a	b	c	
6	3	3	intercepts
$\dfrac{1}{6}$	$\dfrac{1}{3}$	$\dfrac{1}{3}$	reciprocals
1	2	2	clear fractions

Hence, the Miller indices are (122).

(iii)

a	b	c	
2	-3	-3	intercepts
$\dfrac{1}{2}$	$-\dfrac{1}{3}$	$-\dfrac{1}{3}$	reciprocals
3	-2	-2	clear fractions

Hence, the Miller indices are ($3\,\bar{2}\,\bar{2}$).

The negative sign in Miller indices is indicated by placing a bar on the integer.

Example 1.9 : *How many atoms per mm² surface area are there in (i) (110) plane and (ii) (111) plane for copper which has FCC structure and a lattice constant a = 3.61 ×10⁻¹⁰ m ?*

Solution : (i) In Fig. 1.49 (a), ABCD is (110) plane. It contains four quarter atoms and two half atoms.

$$\text{Total number of atoms} = 4\left(\frac{1}{4}\right) + 2\left(\frac{1}{2}\right) = 1 + 1 = 2$$

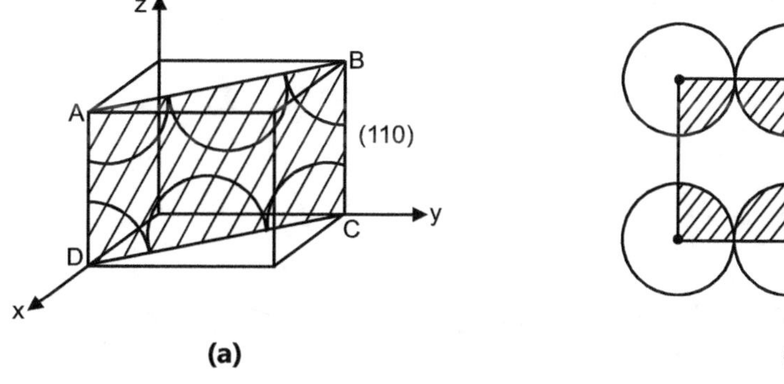

Fig. 1.49

ABCD (110) plane,

$$\text{Area of plane} = BC \times AB = a \times \sqrt{2}\,a = \sqrt{2}\,a^2$$
$$= \sqrt{2} \times (3.61 \times 10^{-7})^2 \text{ mm}^2$$

$$\therefore \quad \text{Number of atoms per mm}^2 = \frac{2}{\sqrt{2}\,(3.61 \times 10^{-7})^2}$$

$$= 1.08 \times 10^{13} \text{ atoms/mm}^2$$

(ii) In Fig. 1.50 (a), ABC is (111) plane. It is the equilateral triangle having each side equal to the face diagonal of the cube.

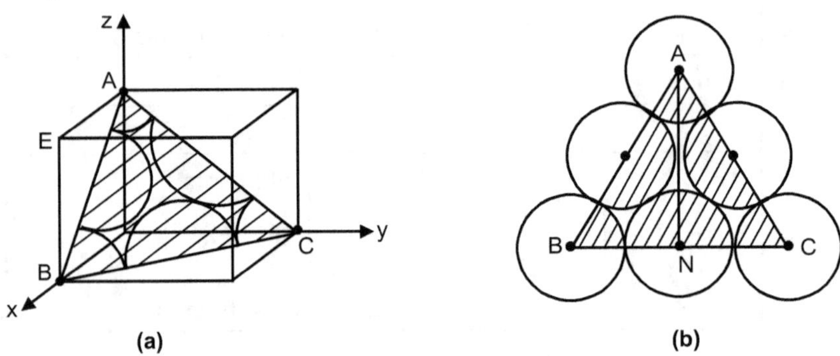

(a)   (b)

Fig. 1.50 : ABC (111) plane

Here, $AB = BC = AC = \sqrt{2}\ a$

$$AN = \frac{\sqrt{3}}{2}\,AC$$

$$= \frac{\sqrt{3}}{2} \cdot \sqrt{2}\,a$$

$$\text{Area of } \triangle ABC = \frac{1}{2}\,BC \times AN$$

$$= \frac{1}{2}\,\sqrt{2}\,a \times \frac{\sqrt{3}}{2}\,\sqrt{2}\,a$$

$$= \frac{\sqrt{3}}{2}\,a^2$$

$$= 0.8660\,a^2$$

From Fig. 1.50 (b), the plane ABC contains three one-sixth of an atom and three one-half.

$$\text{Number of atoms} = 3\left(\frac{1}{6}\right) + 3\left(\frac{1}{2}\right) = 2$$

$$\text{Number of atoms/mm}^2 = \frac{2}{0.8660 \times (3.61 \times 10^{-7})^2}$$

$$= \mathbf{1.778 \times 10^{13} \ atoms/mm^2} \qquad \textbf{... Ans.}$$

Example 1.10 : *In a unit cell of simple cubic structure, find the angle between the normals to pair of planes whose Miller indices are (i) (100) and (010), (ii) (121) and (111).* **(Oct. 17, 15)**

Solution : The direction of two normals are [100], [010] and [121], [111] respectively.

The angle θ between two directions $[u_1 \, v_1 \, w_1]$ and $[u_2 \, v_2 \, w_2]$ is given by,

$$\cos\theta = \frac{u_1 u_2 + v_1 v_2 + w_1 w_2}{(u_1^2 + v_1^2 + w_1^2)^{1/2} (u_2^2 + v_2^2 + w_2^2)^{1/2}}$$

(i)
$$\cos\theta = \frac{1 \times 0 + 0 \times 1 + 0 \times 0}{\sqrt{1^2 + 0^2 + 0^2} \, \sqrt{0^2 + 1^2 + 0^2}} = 0$$

$\therefore \qquad \theta = \mathbf{90°}$ ... **Ans.**

(ii)
$$\cos\theta = \frac{1 \times 1 + 2 \times 1 + 1 \times 1}{\sqrt{1^2 + 2^2 + 1^2} + \sqrt{1^2 + 1^2 + 1^2}} = \frac{4}{\sqrt{6}\sqrt{3}} = \frac{4}{\sqrt{18}}$$

$$\cos\theta = 0.9428$$

$\therefore \qquad \theta = \mathbf{19.47°}$ ... **Ans.**

Example 1.11 : *Show that FCC lattice is the reciprocal of the BCC lattice.*

Solution : Let $\vec{a'}$, $\vec{b'}$ and $\vec{c'}$ be the primitive translation vectors of the body centred cubic lattice. If a is the side of the cube and $\hat{x}$, $\hat{y}$ and $\hat{z}$ are orthogonal unit vectors parallel to the cube edges then,

$$\vec{a'} = \frac{a}{2}(\hat{x} + \hat{y} - \hat{z})$$

$$\vec{b'} = \frac{a}{2}(-\hat{x} + \hat{y} + \hat{z}) \qquad \qquad ...(1)$$

and
$$\vec{c'} = \frac{a}{2}(\hat{x} - \hat{y} + \hat{z})$$

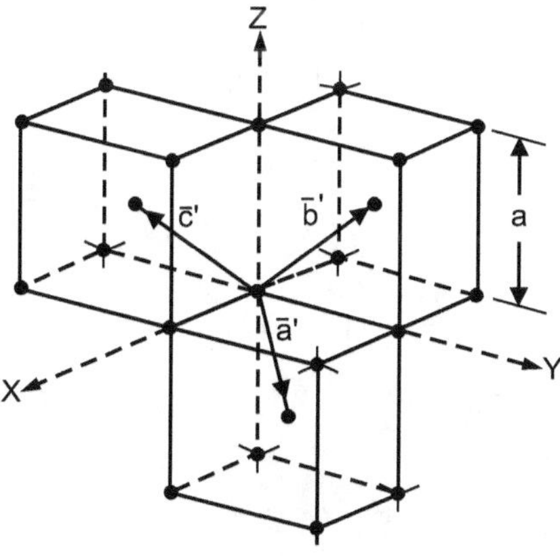

Fig. 1.51

The volume of primitive cell is given by,

$$V = \vec{a}' \cdot (\vec{b}' \times \vec{c}') \qquad \ldots (2)$$

But

$$\vec{b}' \times \vec{c}' = \frac{a^2}{4}(-\hat{x} + \hat{y} + \hat{z}) \times (\hat{x} - \hat{y} + \hat{z})$$

$$= \frac{a^2}{4} \begin{vmatrix} \hat{x} & \hat{y} & \hat{z} \\ -1 & 1 & 1 \\ 1 & -1 & 1 \end{vmatrix}$$

$$= \frac{a^2}{4}\{\hat{x}(1+1) - \hat{y}(-1-1) + \hat{z}(1-1)\}$$

$$\vec{b}' \times \vec{c}' = \frac{a^2}{4}(\hat{x} + \hat{y})$$

$$= \frac{a^2}{2}(\hat{x} + \hat{y}) \qquad \ldots (3)$$

Using equation (3) in (2), we get

$$V = \frac{a}{2}(\hat{x} + \hat{y} - \hat{z}) \cdot \frac{a^2}{2}(\hat{x} + \hat{y}) = \frac{a^3}{4}(2)$$

$$V = \frac{a^3}{2} \qquad \ldots (4)$$

The reciprocal lattice vectors are given by,

$$\vec{A} = \frac{\vec{b}' \times \vec{c}'}{\vec{a}' \cdot (\vec{b}' \times \vec{c}')}, \; \vec{B} = \frac{\vec{c}' \times \vec{a}'}{\vec{a}' \cdot (\vec{b}' \times \vec{c}')},$$

$$\vec{C}' = \frac{\vec{a}' \times \vec{b}'}{\vec{a}' \cdot (\vec{b}' \times \vec{c}')}$$

Now, consider

$$\vec{A} = \frac{\vec{b}' \times \vec{c}'}{\vec{a}' \cdot (\vec{b}' \times \vec{c}')}$$

But

$$\vec{b}' \times \vec{c}' = \frac{a^2}{2}(\hat{x} + \hat{y})$$

and

$$\vec{a}' \cdot (\vec{b}' \times \vec{c}') = \frac{a^3}{2}$$

$$\therefore \qquad \vec{A} = \frac{a^2}{2}\frac{(\hat{x} + \hat{y})}{\frac{a^3}{2}}$$

$$\therefore \qquad \vec{A} = \frac{1}{a} (\hat{x} + \hat{y})$$

Similarly,
$$\vec{B} = \frac{1}{a} (\hat{y} + \hat{z}) \quad \text{and} \quad \vec{C} = \frac{1}{a} (\hat{z} + \hat{x})$$

These vectors are primitive vectors of FCC lattice as shown in Fig. 1.51. Hence, FCC lattice is the reciprocal of the BCC lattice.

Example 1.12 : *Show that the BCC lattice is the reciprocal lattice of FCC lattice.*

Solution : Let $\vec{a}'$, $\vec{b}'$ and $\vec{c}'$ be the primitive vectors of face centred cubic lattice. If a is the side of the cube then $\vec{a}'$, $\vec{b}'$ and $\vec{c}'$ will be,

$$\left. \begin{array}{l} \vec{a}' = \dfrac{a}{2} (\hat{x} + \hat{y}) \\[2mm] \vec{b}' = \dfrac{a}{2} (\hat{y} + \hat{z}) \\[2mm] \vec{c}' = \dfrac{a}{2} (\hat{x} + \hat{z}) \end{array} \right\} \qquad \ldots (1)$$

and

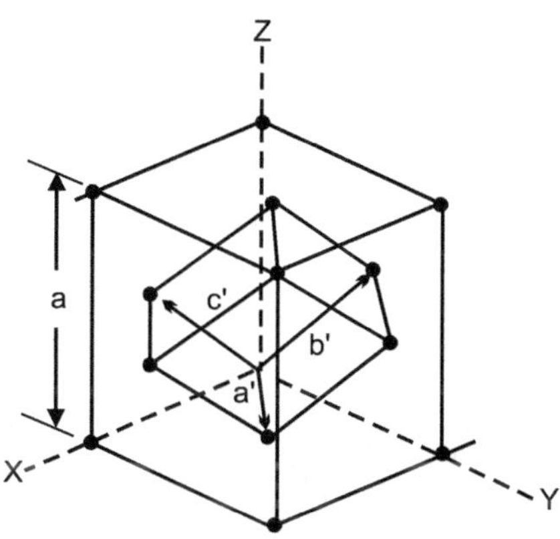

Fig. 1.52

The volume of primitive cell be,

$$V = \vec{a}' \cdot \vec{b}' \times \vec{c}' \qquad \ldots (2)$$

$$\therefore \qquad \vec{b}' \times \vec{c}' = \frac{a}{2} (\hat{y} + \hat{z}) \times \frac{a}{2} (\hat{z} + \hat{x})$$

$$= \frac{a^2}{4} \begin{vmatrix} \hat{x} & \hat{y} & \hat{z} \\ 0 & 1 & 1 \\ 1 & 0 & 1 \end{vmatrix}$$

$$= \frac{a^2}{4} \{\hat{x}(1-0) - \hat{y}(0-1) + \hat{z}(0-1)\}$$

$$= \frac{a^2}{4}(\hat{x} + \hat{y} - \hat{z}) \qquad \ldots (3)$$

Now,

$$V = \frac{a}{2}(\hat{y} + \hat{x}) \cdot \frac{a^2}{4}(\hat{x} + \hat{y} - \hat{z})$$

$$= \frac{a^3}{8}(1 + 1)$$

$$V = \frac{a^3}{4} \qquad \ldots (4)$$

The primitive translation vectors of the reciprocal of FCC lattice are given by,

$$\vec{A} = \frac{\vec{b'} \times \vec{c'}}{\vec{a'} \cdot (\vec{b'} \times \vec{c'})}, \quad \vec{B} = \frac{\vec{c'} \times \vec{a'}}{\vec{a'} \cdot (\vec{b'} \times \vec{c'})}, \quad \vec{C} = \frac{\vec{a'} \times \vec{b'}}{\vec{a'} \cdot (\vec{b'} \times \vec{c'})}$$

Now, consider

$$\vec{A} = \frac{\vec{b'} \times \vec{c'}}{\vec{a'} \cdot (\vec{b'} \times \vec{c'})}$$

But

$$\vec{b'} \times \vec{c'} = \frac{a^2}{4}(\hat{x} + \hat{y} - \hat{z})$$

and

$$\vec{a'} \cdot \vec{b'} \times \vec{c'} = \frac{a^3}{4}$$

$\therefore$

$$\vec{A} = \frac{a^2}{4} \frac{(\hat{x} + \hat{y} - \hat{z})}{\dfrac{a^3}{4}}$$

$$\vec{A} = \frac{1}{a}(\hat{x} + \hat{y} - \hat{z})$$

Similarly,

$$\vec{B} = \frac{1}{a}(-\hat{x} + \hat{y} + \hat{z})$$

and

$$\vec{C} = \frac{1}{a}(\hat{x} - \hat{y} + \hat{z})$$

These vectors $\vec{A}$, $\vec{B}$, $\vec{C}$ are primitive vectors of BCC lattice. Hence, BCC lattice is the reciprocal lattice of FCC lattice.

Summary

1. A crystalline solid is a substance whose constituent particles possess a regular orderly arrangement

2. An amorphous solid is a substance whose constituent particles do not possess a regular orderly arrangement.

3. An arrangement of infinite number of imaginary points in space with each point having identical surrounding is known as crystal lattice.

4. The structure unit of a crystal is called basis which may consist of an atom, an ion or a molecule.

5. Pair of translation vectors such that the elementary parallelogram formed by them is of minimum area. Such translation vectors are called primitive or fundamental translation vectors.

6. The unit cell whose concurrent sides qualify to be the primitive translation vectors is called primitive cell.

7. There exist certain operations which when operated on a given structure, the structure remains unchanged or invariant. Such operations are called symmetry operations and the structure is said to possess symmetry under that operation.

8. It was shown by Bravais that only 14 different networks of lattices can be generated in which lattice points arranged in 3-dimensional space of every network are such that each point has identical surroundings. These are known as Bravais lattices.

9. According to Miller, it is more useful to describe the orientation of a plane by the reciprocal of its numerical parameters rather than by its linear parameter. Such parameters are called Miller Indices.

10. Interplaner distance is the spacing between two parallel planes in a given cell.

    Expression for interplaner distance $d = \dfrac{1}{\sqrt{\dfrac{h^2}{a^2} + \dfrac{k^2}{b^2} + \dfrac{l^2}{c^2}}}$

11. The atomic radius is defined as half the distance between two nearest neighbouring atoms of same kind.

12. The total number of atoms per unit cell is given by,

    $$N = N_i + \frac{N_f}{2} + \frac{N_c}{8}$$

13. The number of nearest neighbours that an atom has in a unit cell is called as co-ordination number.

14. Density of crystal material is the ratio of mass of unit cell to the volume of unit cell.

15. Packing fraction is defined as the ratio of volume of atoms occupying the unit cell to the total volume of the unit cell relating to that structure.

16. Reciprocal lattice is defined as a vector having magnitude equal to the reciprocal of the interplaner spacing d_{dkl} and direction coinciding along normal to the (hk$l$) planes.

$$|\vec{\sigma}_{hkl}| = \frac{1}{d_{hkl}}$$

Exercise

(A) Short Answer Type Questions :

1. Define the following terms used in crystal structure :
    (i)   space lattice,
    (ii)   basis
    (ii)   unit cell,
    (iii)   primitive cell,
    (iv)   co-ordination number,
    (v)   packing fraction.
    (vi)   fold number
    (vii)   primitive translational vectors.

2. Show that for simple cubic system, r = a/2.

3. Obtain number of atoms per unit cell for simple cubic crystal.

4. Obtain number of atoms per unit cell for FCC crystal.

5. Obtain number of atoms per unit cell for BCC crystal.

6. Give co-ordination number for SC, FCC and BCC structure.

7. Draw NaCl structures.

8. Distinguish between crystalline solid and amorphous solid.

9. What do you mean by primitive translational vectors and non-primitive translational vectors?

10. Define the term symmetry operations.

11. Define the term reciprocal lattice vector.

12. What do you understand by fold number ? How it is calculated ?

(B) Long Answer Type Questions :

1. What is a Bravais lattice? What are different space lattices in the cubic system ?

2. What do you mean by space lattice? Explain space lattice with two-dimensional square array of points.

3   Show that a five fold rotation axis does not exist in crystal lattice.

4.   What are Miller indices of the plane? How they are determined?

5.   Obtain an expression for interplaner distance. Hence, show that for simple cubic system,

$$d_{hkl} = \frac{a}{\sqrt{h^2 + k^2 + l^2}}$$

6.   What is crystal system? Describe briefly the seven systems of crystals with suitable diagrams.

7.   With proper diagrams explain BCC, FCC and HCP structures in detail.

8.   What are symmetry operations? Explain three principle types of symmetry operation.

9.   What is co-ordination number? Write co-ordination number for simple cubic, BCC and FCC structures.

10. Explain the term packing fraction. Show that packing fraction for simple cubic, BCC and FCC structures are 0.52, 0.68 and 0.74 respectively.

11. Discuss crystal structures of NaCl, CsCl, ZnS and hcp in detail.

12. What is reciprocal lattice? Obtain expression for reciprocal lattice vectors $\bar{A}$, $\bar{B}$ and $\bar{C}$.

13. Show that reciprocal of the reciprocal lattice is direct lattice.

14. Consider a plane hkl in a crystal plane, prove that the reciprocal lattice vector

      $G = hb_1 + kb_2 + lb_3$ is perpendicular to this plane.

15. What are (100), (110) and (111) planes? Explain and draw sketches.

16. Explain the procedure to generate reciprocal lattice point in the space.

17. What is reciprocal lattice vector? Show that for two-dimensional square lattice,

$$\sigma = \frac{\sqrt{h^2 a^2 + k^2 b^2}}{ab}$$

18. With suitable diagram, explain diamond cubic structure.

19. For BCC structure, show that $r = \dfrac{\sqrt{3}}{4} a$.

(C) Unsolved Problems :

1.   FCC structure has atomic radius of 1.75 A°. Find the spacing of (i) 100 plane, (ii) 200 plane.   **(Ans.** 4.96 A°, 2.48 A°)

2. Calculate lattice constant of NaCl crystal using weight of NaCl = 58.45, density of NaCl = 2170 kg/m³. Avogadro's number = 6×10^{23}. **(Ans.** a = 5.64 A°)

3. Show that for simple cubic lattice : $d_{100} : d_{110} : d_{111} = \sqrt{6} : \sqrt{3} : \sqrt{2}$

4. How many atoms are assigned to the unit cell of (a) primitive, (b) FCC and (c) BCC lattice? **(Ans.** 1, 4, 2)

5. Calculate the Miller indices of crystal planes which cut through the crystal axes at (i) (2a, 3b, c), (ii) (a, b, c) and (iii) (2a, − 3b, − 3c).

(Ans. (i) 3 2 6, (ii) 1 1 1, (iii) $3\ \bar{2}\ \bar{2}$)

6. Find the atomic spacing for a crystal of rock salt, NaCl, whose formula mass is 58.5 u and whose density is 2.16×10^3 kg/m³. Where, 1 u = amu = 1.66×10^{-27} kg.

Hint : For NaCl, k = 2

$$\left(\textbf{Ans. } d = \left[\frac{M}{K\rho} \times 1.66 \times 10^{-27} \text{ kg/u}\right]^{1/3}\right)$$

7. Calculate number of atoms per unit cell of a metal having lattice constant 0.29 nm and density 7870 kg/m³. Atomic weight of metal is 55.85. **(Ans** 2)

8. Copper has fcc structure and the atomic radius is 0.1278 nm. Calculate the interplaner spacing for (1 1 1) and (3 2 1) planes. **(Ans.** 0.362 nm and 0.097 nm)

Chapter **2**...

X-Ray Diffraction and Other Characterization Techniques

Contents ...

(A) X-RAY DIFFRACTION

2.1 Introduction

Sir William Lawrence Bragg (31 March 1890 – 1 July 1971) was an Australian-born British physicist and X-ray crystallographer, discoverer (1912) of the Bragg law of X-ray diffraction, which is basic for the determination of crystal structure. He was joint winner (with his father, William Henry Bragg) of the Nobel Prize in Physics in 1915.

- X-rays are electromagnetic waves like ordinary waves and hence show interference and diffraction. The wavelength of X-rays is of the order of 1 A° (or 0.1 nm), hence ordinary devices such as ruled diffraction grating do not produce noticeable effects with X- ray. In 1912, German scientist Laue suggested that a crystal consisting of 3 D arrangement of regularly spaced atoms could serve the purpose of grating.

- Laue's associates Friedrich and Knipping successfully showed X-ray diffraction by passing them through a thin crystal of zinc blende. Observed diffraction pattern consisted of a central spot and series of spot arranged in a definite pattern around central spot. Such symmetrical pattern is known as Laue pattern. The interpretation of such pattern was given by W. L. Bragg.

2.2 Bragg's Diffraction

- When a beam of monochromatic X-rays falls on a crystal, it is scattered by individual atoms which are arranged in sets of parallel planes. Each atom becomes a source of scattered radiations.

- The combined scattering of X-rays from these planes can be looked upon as reflections from these planes. Because of this, Bragg's scattering is usually referred to as Bragg's reflection and these planes are known as Bragg's planes. It is due to the presence of such sets of parallel planes that a crystal acts as reflection grating.

- For certain incident angles, reflection from these sets of parallel planes are in phase with each other. In order to understand this, consider incoming rays OE and O'A incident at the angle θ to the planes (hk*l*). The rays AP and EP' are scattered at an angle θ from these planes as shown in Fig. 2.1.

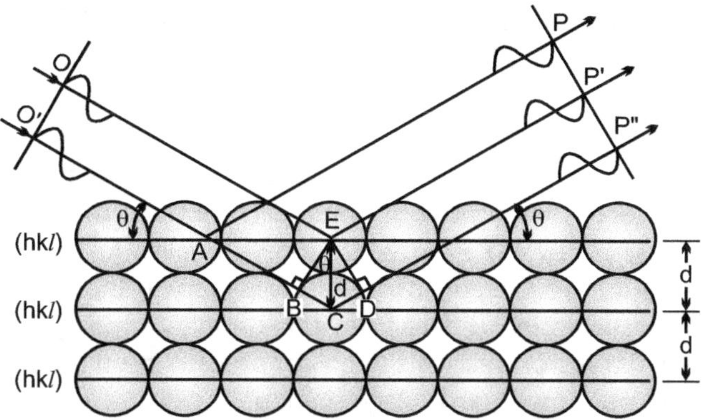

Fig. 2.1 : Diffraction of X-rays

- The total path lengths of the rays O'AP and OEP' are the same, therefore, these rays are said to be scattered in phase with each other. Hence reflected X-rays add up to give the resulting reflection strong. For some other values of θ, the X-ray reflections from different planes are antiphase with each other so that resulting reflection is either zero or extremely minimum as shown in Fig. 2.2 (a) and (b).

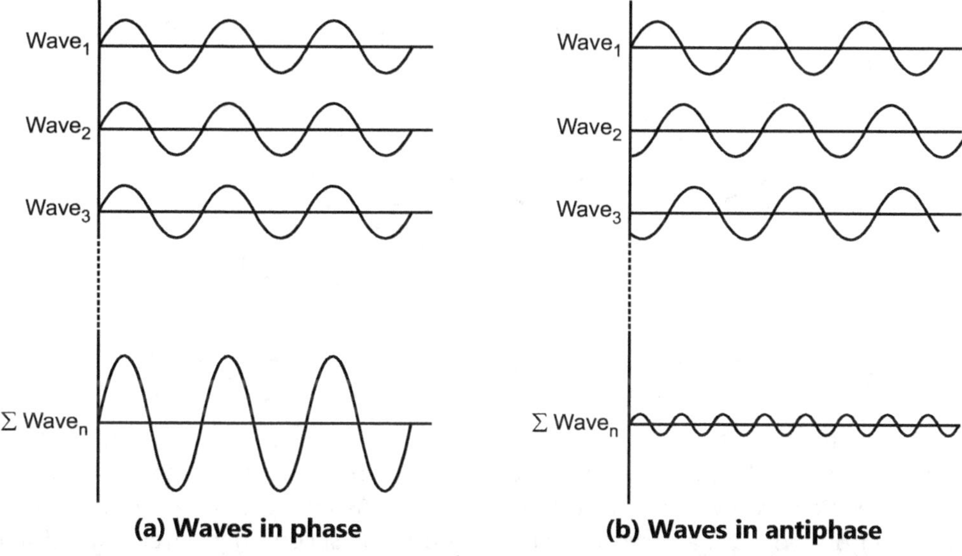

(a) Waves in phase   (b) Waves in antiphase

Fig. 2.2

2.3 Crystal as a Grating   (Oct. 16)

- To show the wave nature of X-rays, it is essential that they should exhibit phenomena like interference and diffraction. These conditions are well known for light waves. From the theory of grating we know that the spacing between the lines ruled on grating should be of the order of the wavelength of light to obtain the diffraction pattern. For example, when yellow light of wavelength $\lambda \sim 6000\ \text{A}°$ is passed through a transmission grating having 6000 lines per cm, a diffraction pattern is observed.

- As X-rays are of considerably shorter wavelengths ($\lambda \approx 1$ A°), the diffraction grating should have about 40 million lines per cm so that using such type of grating one can observe diffraction pattern. It is practically impossible to have such an artificial transmission grating.

- Laue suggested that a crystal which consists of three-dimensional array of regularly spaced atoms can serve the purpose of grating. The interatomic separations in a crystal are of the order of a few A°. Hence, the X-rays qualify to be the best to undergo diffraction because the wavelength of the X-ray is also of the order of few angstroms. The crystal used as a grating differs from the ordinary grating in the sense that the diffracting centres (atoms or ions) in the crystal are not in one plane. Hence the crystal acts as a space grating rather than a plane grating.

2.4 Bragg's Law and Bragg's Diffraction Condition   (Oct. 17, 16)

- Consider a beam of X-rays of wavelength λ is incident at a glancing angle θ on a set of parallel planes of the crystal as shown in Fig. 2.3.

- The atomic planes are considered to be semi-transparent, therefore, they allow a part of X-rays to pass through and reflect the other part. The angle of incident θ is known as Bragg's angle and it is equal to angle of reflection.

- The first ray is reflected from atom A in the plane 1, whereas the second ray is reflected from atom B lying in the plane 2 immediately below the atom A.

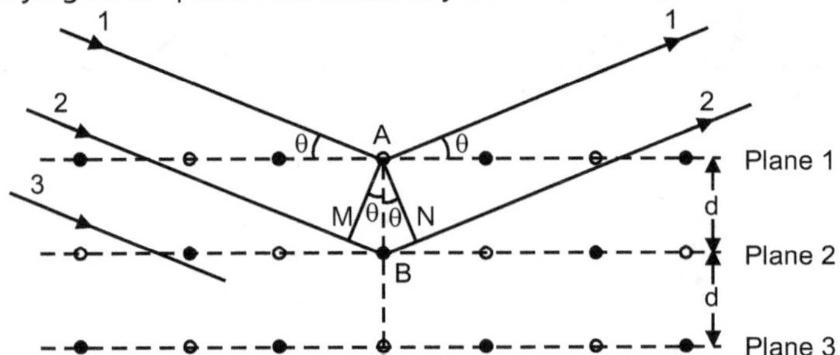

Fig. 2.3 : Reflection of X-rays from lattice planes in a crystal

- The X-rays reflected from the atoms A and B will be in phase or antiphase depending upon their path difference. Let us find the path difference between these two rays. Draw the perpendiculars AM and AN on the second ray. As two rays have travelled the same distance from points A and N onwards, therefore, the second ray travels extra distances MB and BN.

Therefore, the path difference between the two reflected beams is,

$$\Delta = MB + BN \qquad \qquad \text{... (2.1)}$$

But from Fig. 2.3, $BN = MB = d \sin \theta$

$\therefore \qquad \qquad \Delta = d \sin \theta + d \sin \theta = 2d \sin \theta \qquad \qquad \text{... (2.2)}$

where d is the distance between two consecutive planes in the crystal and θ angle between the incident ray and the planes of reflection (glancing angle).

- Two reflected beams will be in phase with each other, if the path difference is integral multiple of λ and will be in antiphase if the path difference is an odd multiple of $\lambda/2$.

- Therefore, the condition for producing maxima will be

$$2d \sin \theta = n\lambda \qquad \qquad \text{... (2.3)}$$

where n = 1, 2, 3, ... etc. for first order, second order, third order, maxima respectively.

- This condition is also known as **Bragg's condition for reflection**. This relation is known as Bragg's law.

- As $\sin \theta$ has maximum value 1, for a typical value of interplaner spacing of 1.5 A°, equation (2.3) gives the upper limit of λ for obtaining first order reflection.

Therefore, substituting n = 1, $\sin \theta$ = 1, d = 1.5 A° in equation (2.3), we get,

$$2 \times 1.5 \, (1) = (1) \, \lambda$$

$$\therefore \qquad \qquad \lambda = 3 \text{ A°}$$

From equation (2.3),

$$\sin \theta = \frac{n\lambda}{2d}$$

For n = 1 and d = 1.5 A°, we get

$$\sin \theta = \frac{\lambda}{3}$$

- When λ is greater than 3 A°, the value of $\sin \theta$ becomes greater than 1 which is not allowed. Hence there will be no reflection if λ is greater than 3 A°.

- The reflection can be obtained for smaller values of λ for those sets of planes that have spacing less than 1.5 A° as well as increasing number of higher order reflection. If wavelength is very small, of the order 0.1 A° of X-rays , then these rays produce other effects such as knocking off electrons from the atoms of the crystal and getting absorbed in the process.

- Bragg's law may be used to find the wavelength (λ) of X-rays if interplaner spacing (d) is known. It can be used for determining the lattice parameters of cubic crystals.

2.5 Bragg's Law in Reciprocal Lattice   (April 17; Oct. 17, 15)

- The diffraction of X-rays occurs from various sets of parallel planes having different orientations and different interplaner distances. In some crystals, there are number of sets of parallel planes with different orientations. It becomes difficult to visualize all such planes due to their two-dimensional nature.

- This problem was simplified by P. P. Ewald by proposing a new type of lattice known as reciprocal lattice. The crystal lattice can be conveniently represented by reciprocal lattice also. The photographs of X-ray diffraction find easier description in terms of reciprocal lattice rather than direct lattice.

- For the construction of reciprocal lattice to a direct lattice, take origin at some arbitrary point and draw normals to every set of parallel planes of the direct lattice. The length of each normal should be equal to the reciprocal of interplaner distance for the corresponding set of parallel planes. The terminal points of all normals form the reciprocal lattice.

- Let us consider a well collimated beam of monochromatic X-rays of frequency v incident on a crystal as shown in Fig. 2.4. The energy of incident X-ray beam = hv or $\hbar\omega$.

- Let the frequency of reflected X-rays be v'. The energy of reflected X-ray beam = hv' or $\hbar\omega'$, where $\omega' = 2\pi v'$

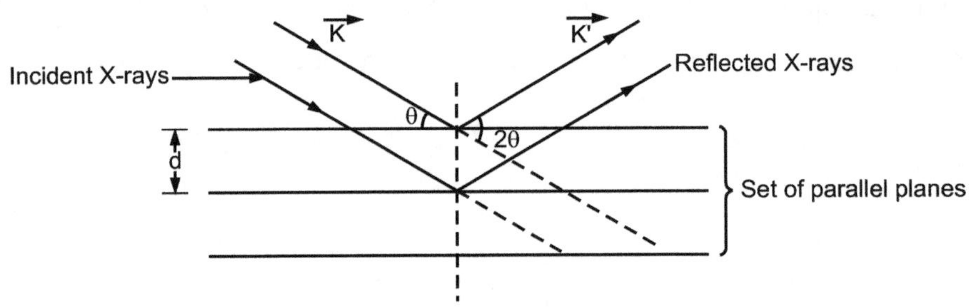

Fig. 2.4 : Bragg's law

For elastic scattering of X-rays,

Incident energy of X-ray = Reflected energy of X-ray

$$\hbar\omega = h\omega'$$

or $$\omega = \omega' \qquad \qquad \qquad ... (2.4)$$

- Let $\vec{K}$ and $\vec{K'}$ be the wave vectors in the direction of propagation of incident and reflected X-ray beam respectively.

- As the space between the parallel planes is free space, the magnitude of the wave vectors is given by

$$|\vec{K}| = \frac{\omega}{c} \quad \text{and} \quad |\vec{K'}| = \frac{\omega'}{c} \qquad \qquad ... (2.5)$$

where c is the velocity of electromagnetic waves or X-rays in free space.

Using equation (2.4) in equation (2.5), we get

$$|\vec{K}| = |\vec{K'}|$$

- Let us assume that the wave vectors $\vec{K}$ and $\vec{K'}$ are related by the equation $\vec{K'} = \vec{K} + \vec{G}$, where $\vec{G}$ is any reciprocal lattice joining tips of $\vec{K}$ and $\vec{K'}$.

We can write, $$\vec{K'} \cdot \vec{K'} = (\vec{K} + \vec{G}) \cdot (\vec{K} + \vec{G})$$

or $$K'^2 = K^2 + G^2 + 2\vec{K} \cdot \vec{G}$$

But, $$|\vec{K}| = |\vec{K'}|$$

$$\therefore \quad 2\vec{K} \cdot \vec{G} + G^2 = 0 \qquad \ldots (2.6)$$

Equation (2.6) is known as **Bragg's diffraction condition in reciprocal lattice**.

2.6 Ewald Construction (April 16)

- We have seen that the incident and reflected X-ray beams are separated by an angular separation of angle 2θ. Consider reciprocal lattice points as shown in Fig. 2.6.

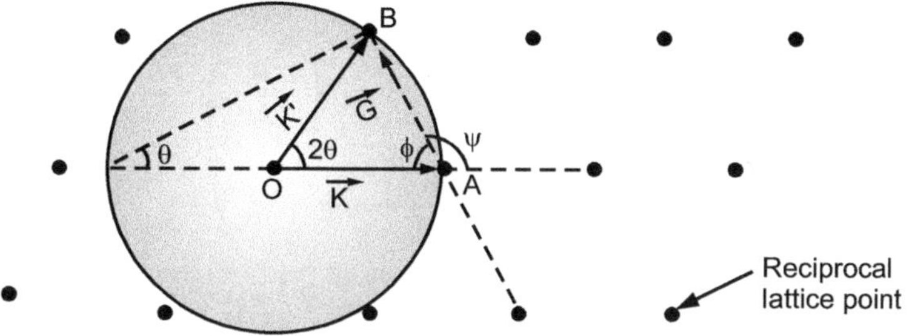

Fig. 2.5 : Ewald construction

- Let us select the direction of propagation of wave such that the incident wave vector $\vec{K}$ terminates on one of the lattice points, say A. Let 'O' be the origin of $\vec{K}$. Now, construct a sphere in a reciprocal lattice space with radius OA and having centre O.

- As the magnitudes of $\vec{K}$ and $\vec{K'}$ are equal, it is obvious that the tip of reflected wave vector $\vec{K'}$ will also lie somewhere on the surface of the sphere. Let the tip of $\vec{K'}$ lie at point B as shown in Fig. 2.5.

- If the tip of $\vec{K}$ and $\vec{K'}$ are joined, we get a vector $\vec{G}$.

$$\therefore \quad \vec{K'} = \vec{K} + \vec{G}, \text{ where } \vec{G} \text{ is a reciprocal lattice vector.}$$

- Let us show that the wave vector $\vec{K'}$ given by above relation qualifies to be a possible Bragg reflected wave vector.

In Fig. 2.5, since seg (OA) $\cong$ seg (OB), the $\triangle$ OAB is isosceles triangle.

$$\therefore \quad \angle OAB \cong \angle OBA \text{ (say } \phi)$$

In $\triangle$ OAB, $$2\theta + 2\phi = \pi$$

or $$\phi = \left(\frac{\pi}{2} - \theta\right)$$

The angle between positive directions of $\vec{K}$ and $\vec{G}$ is

$$\psi = \pi - \phi = \pi - \left(\frac{\pi}{2} - \theta\right)$$

$$\therefore \qquad \psi = \frac{\pi}{2} + \theta$$

Now, $\qquad \vec{K} \cdot \vec{G} = KG \cos \psi = KG \cos \left(\frac{\pi}{2} + \theta\right)$

But, $\qquad \cos \left(\frac{\pi}{2} + \theta\right) = - \sin \theta$

$$\therefore \qquad \vec{K} \cdot \vec{G} = - KG \sin \theta$$

Using above equation in Bragg's diffraction condition in reciprocal lattice $(2\vec{K} \cdot \vec{G} + G^2 = 0)$, we get

$$- 2KG \sin \theta + G^2 = 0$$

or $\qquad G = 2K \sin \theta \qquad\qquad (\because G \neq 0) \quad \dots (2.7)$

where G is a reciprocal lattice vector.

The reciprocal lattice vector $\vec{G}$ is always expressed as

$$G = n\frac{2\pi}{d} \qquad\qquad \dots (2.8)$$

where n is integer and d is the interplaner distance in the direct lattice.

Substituting equation (2.8) in equation (2.7), we get

$$n\frac{2\pi}{d} = 2K \sin \theta$$

But, $\qquad K = \frac{2\pi}{\lambda}$

$$\therefore \qquad n\frac{2\pi}{d} = 2\frac{2\pi}{\lambda} \sin \theta$$

or $\qquad d \sin \theta = n\lambda$

which is Bragg's condition in direct lattice.

Thus, Bragg's diffraction condition in reciprocal lattice is exactly equivalent to the Bragg's condition in the direct lattice.

2.7 Experimental Methods of X-Ray Diffraction

- We can determine the crystal structure using standard methods of X-ray crystallography. These methods are described below.

  (1) Laue method: In this method, a single crystal is held stationary and a beam of white X-ray radiation of continuous wavelength (0.2 A° to 2 A°) is inclined on it at a fixed glancing angle θ. The crystal selects and diffracts the discrete values of wavelength (λ) satisfying the Bragg's condition $2d \sin \theta = n\lambda$.

  Laue in his very first experiments used white radiation of all possible wavelengths and allowed this radiation to fall on a stationary crystal. The crystal diffracted the X-ray beam and produced a very beautiful pattern of spots which conformed exactly with the internal symmetry of the crystal. Let us analyse the experiment with the aid of the Bragg equation. The crystal was fixed in position relative to the X-ray beam, thus not only was the value for d fixed, but the value of θ was also fixed (Refer Fig. 2.6).

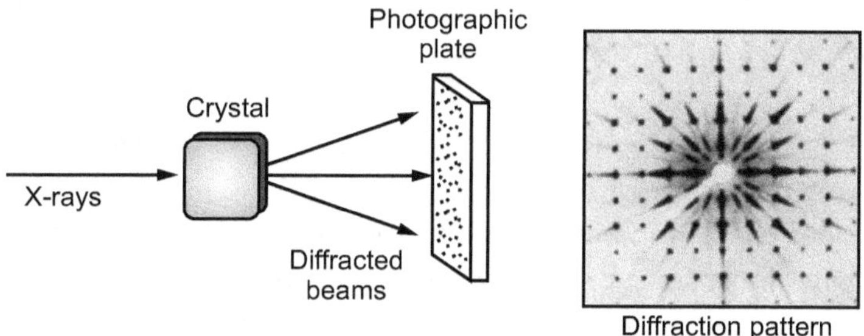

Fig. 2.6 : Transmission Laue camera

The only possible variables therefore are the integer n and wavelength λ, with the result that in the Laue photograph each observed reflection corresponds to the first order of reflection of a certain wavelength, the second order of half the wavelength and a third order of a third of the wavelength, and so on. The Laue photograph which is obtained from a single crystal is simply a stereographic projection of the planes of the crystal.

  (2) Rotating crystal method : In this method, a monochromatic beam of X-rays is incident on a single crystal rotating about a fixed axis. The variation in the angle θ brings different atomic planes into the position for reflection satisfying Bragg's condition $2d \sin \theta = n\lambda$ each producing spot on the film. The position on the film when developed indicates orientation of the crystal at which spot is formed. The data obtained from these spots revels information about structures of ordinary and complex molecules.

In the rotating-crystal method a single crystal is rotated about a fixed axis in a beam of monoenergetic X-rays. Due to rotation, there is variation in the angle θ. The variation in the angle θ brings different atomic planes into position for reflection. A modified rotating-crystal method is used for structure determination when a single-crystal specimen is available.

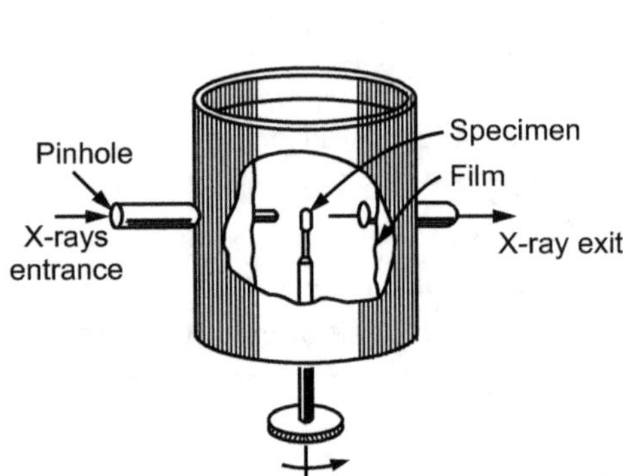

<table>
<tr><td>

Fig. 2.7 (a) : A camera used in the rotating crystal method

</td><td>

Fig. 2.7 (b) : Intensity distribution from a radiation of a 30 keV X-ray with molybdenum

</td></tr>
</table>

- A simple rotating-crystal X-ray camera is shown in Fig. 2.7 (a). The film is mounted in a cylindrical holder concentric with a rotating spindle on which the single crystal specimen is mounted. The dimensions of the crystal usually need not be greater than 1 mm. The incident X-ray beam is made nearly monochromatic by a filter or by reflection from an earlier crystal. The beam is diffracted from a given crystal plane whenever in the course of rotation the value of θ satisfies the Bragg equation. Beams from all planes parallel to the vertical rotation axis will lie in the horizontal plane. Planes with other orientations will reflect in layers above and below the horizontal plane.

- The intensity distribution of the radiation from a 30 keV X-ray tube with a molybdenum target is shown by Fig. 2.7 (b).

    (3) Debye-Scherrer method (Powder method): Laue method and rotating crystal method can be applied only if, single crystals of reasonable size are available. However, most of the crystalline substances naturally available are in the form of polycrystals. The method of preparation of single crystal is quite difficult.

Debye and Scherrer method provides the details about the crystal structure even if the specimen taken for investigation is in the form of polycrystals.

The crystalline substance be finely grounded and powdered so that its tiny crystals are randomly oriented. The powder specimen C stuck on hair by means of gum and placed in the path of a narrow monochromatic X-rays as shown in Fig. 2.8 (a).

Basic principle of this method is that millions of tiny crystals in the powder are randomly oriented. All possible diffraction planes will be available for Bragg reflection. Such reflection will take place from many sets of parallel planes lying at different angles to the incident X-ray beam. Moreover, each set will give not only first order reflections but those of higher orders as well. Since all orientations are equal likely, the reflected rays will form a cone whose axis lies along the direction of incident beam and whose semi-vertical angle is twice the glancing angle for that particular set of planes which is shown in Fig. 2.8 (b).

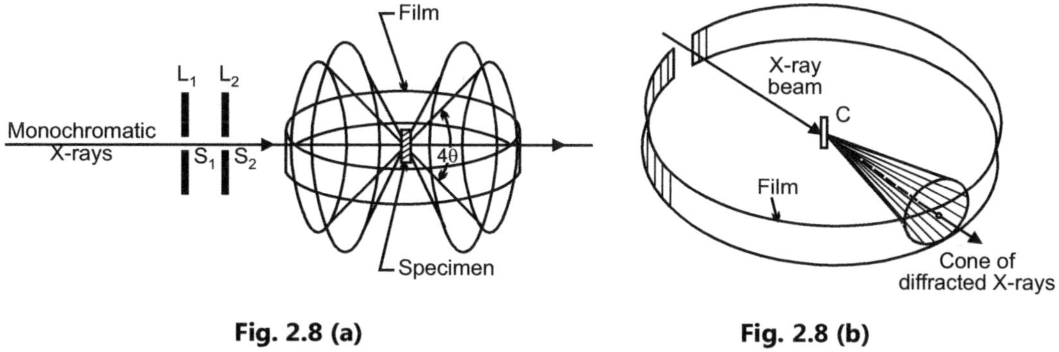

Fig. 2.8 (a)   **Fig. 2.8 (b)**

For each set of planes and for each order, there will be such cone reflected X-rays. Their interactions with a photographic film form a series of concentric circular rings. Radii of these rings are recorded and the film can be used to find the glancing angle and hence interplaner spacing of crystalline substance.

Experimental arrangement: The X-rays from the tube are passed through filter F, it gives out monochromatic X-rays collimated by two fine slits S_1 and S_2 .Then X-ray beam falls on the powder specimen C stuck on hair by means of gum as shown in Fig. 2.9 (a) and suspended along the axis of the cylindrical camera called Debye-Scherrer camera.

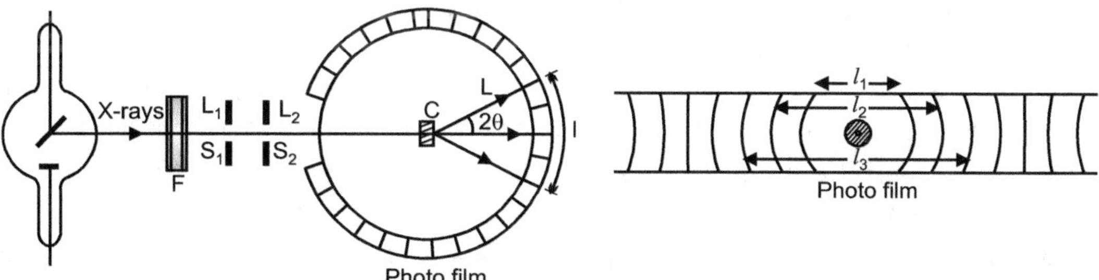

Fig. 2.9 (a) : Powdered crystal spectrometer   Fig. 2.9 (b) : Developed film

The photographic film is mounted round the inner surface of camera and almost completely surrounds C in order to receive X-rays diffracted up to 180°. Since the film is of narrow width, only parts of the circular rings register on the film as shown in Fig. 2.9 (b).

The curvature of these arcs reverses after the angle of diffraction exceeds 90°. If the distance between symmetrical lines on the stretched photograph are l_1, l_2, l_3, ..., etc. and diameter of the cylindrical film is D then,

$$\frac{l_1}{\pi D} = \frac{4\theta_1}{360°}$$

$$\therefore \qquad \theta_1 = \frac{90°}{\pi D} \, l_1$$

$$\therefore \qquad \theta_1 = k l_1 \qquad \text{where, } k = \frac{90°}{\pi D}$$

Similarly, $\theta_2 = k l_2$ and $\theta_3 = k l_3$, etc. Using these values of θ in Bragg's formulae, interplaner spacing d can be calculated as follows.

We have,

$$2d \sin \theta = n\lambda \qquad \text{where n is constant.}$$

Differentiating, we get,

$$2 \Delta d \sin \theta - 2d \cos \theta \, \Delta\theta = 0$$

$$\therefore \qquad \frac{\Delta\theta}{\Delta d} = -\frac{\tan \theta}{d}$$

The radius $\dfrac{\Delta\theta}{\Delta d}$ will be maximum when $\theta = 90°$ or $2\theta = 180°$.

This means X-rays are reflected back along initial path. This is known as the **back reflection.** Such back reflections cannot be recorded experimentally. Hence $\dfrac{\Delta\theta}{\Delta d}$ should be large.

The advantage of this method is that it does not require large single crystals and almost any substance can be grinded into powder.

By measuring the distances between symmetrical lines and line intensities, following information can be obtained.

1. A crystalline substance can be distinguished from an amorphous one because crystalline substance produces lines on a photograph, while amorphous does not.

2. If photographs are made before and after thermal or mechanical treatment of a crystalline substance. Their comparison can reveal any alteration in the lattice.

3. With the aid of standard photographs of known chemical compounds their presence may be detected in an unknown mixture.

4. The size of unit cell and the type of the lattice can be determined in simple cases.

The powder method has been employed in the study of microscopic substance like metals, alloys, carbon, fluorescent powders and other forms where single crystals are not available.

2.8 Analysis of Cubic Structure by Powder Method

- Consider a beam of monochromatic X-rays of wavelength λ incident on the powdered specimen of cubic crystal. Let the pattern recorded on the photographic film be as shown in Fig. 2.9 (b). Let l_1, l_2 and l_3 be the distances between symmetrical lines on the stretched film and θ_1, θ_2 and θ_3 be the corresponding angles of diffraction respectively.

- The circular symmetrical lines on the film can be identified by an experimental parameter ϕ related to θ (angle of diffraction) by the relation $2\theta + \phi = \pi$ (Refer Fig. 2.10).

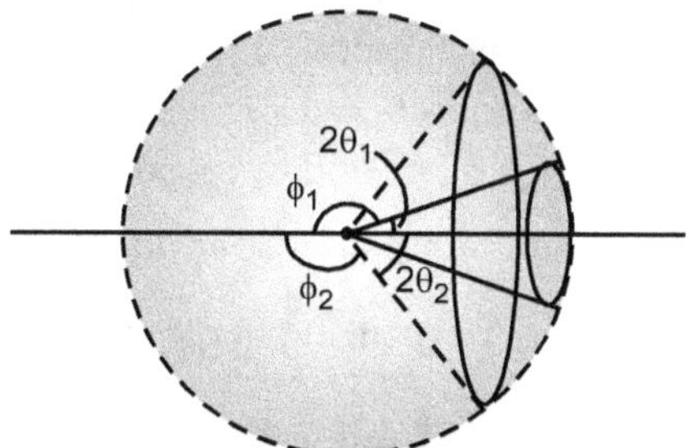

Fig. 2.10 : Relation between θ and ϕ

- We know that the interplaner distance between two parallel planes having Miller indices (hk$l$) for cubic crystal is given by the relation

$$d_{(hkl)} = \frac{a}{\sqrt{h^2 + k^2 + l^2}} \qquad \ldots (2.9)$$

where a is the lattice spacing.

The Bragg's diffraction condition is

$$2d \sin \theta = n\lambda$$

or

$$d = \frac{n\lambda}{2 \sin \theta} \qquad \ldots (2.10)$$

Equating equations (2.10) and (2.9), we get

$$\frac{n\lambda}{2 \sin \theta} = \frac{a}{\sqrt{h^2 + k^2 + l^2}}$$

Squaring both sides of above equation and rearranging the terms, we get

$$n^2\lambda^2 (h^2 + k^2 + l^2) = 4a^2 \sin^2 \theta$$

Since n, h, k and l are integers, $n^2 (h^2 + k^2 + l^2)$ must be a positive integer, say N.

$$\therefore \qquad N\lambda^2 = 4a^2 \sin^2 \theta$$

or

$$\frac{\sin^2 \theta}{N} = \frac{\lambda^2}{4a^2}$$

But, λ and a are constants in this case.

$$\therefore \quad \frac{\sin^2\theta}{N} = \text{constant}$$

For a simple cubic crystal, it is found that [using equation (2.9)]

$$d_{100} : d_{110} : d_{111} = 1 : \frac{1}{\sqrt{2}} : \frac{1}{\sqrt{3}}$$

- The above relation can be verified for planes 100, 110 and 111 by taking specific experimental data of cubic crystal. Suppose for $\lambda = 1.3$ A°, the maxima of reflected X-rays are obtained at glancing angles $\theta_1 = 19.2°$, $\theta_2 = 27.72°$ and $\theta_3 = 34.73°$ for parallel planes having interplaner spacings d_1, d_2 and d_3 respectively.

- To find the lattice constant 'a' and interplaner spacing d for a set of parallel planes, let us prepare a table for $\left(\dfrac{\sin^2\theta}{N}\right)$ values for glancing angles θ_1, θ_2 and θ_3.

Table 2.1

↓ Values of N		$\theta_1 = 19.2°$ $\sin\theta_1 = 0.3289$ $\sin^2\theta_1 = 0.1082$	$\theta_2 = 27.72°$ $\sin\theta_2 = 0.4651$ $\sin^2\theta_2 = 0.2163$	$\theta_3 = 34.73°$ $\sin\theta_3 = 0.5697$ $\sin^2\theta_3 = 0.3246$
2	$\dfrac{\sin^2\theta}{N}$	**0.0541**	0.1081	0.1623
3		0.0361	0.0721	0.1082
4		0.0271	**0.0541**	0.0812
5	Values	0.0216	0.0433	0.0649
6	→	0.0180	0.0361	**0.0541**

From Table 2.1, constant value of $\dfrac{\sin^2\theta}{N} = 0.0541$.

But $$\frac{\sin^2\theta}{N} = \frac{\lambda^2}{4a^2}$$

$$\therefore \quad a^2 = \frac{\lambda^2}{4 \times \dfrac{\sin^2\theta}{N}} = \frac{(1.3)^2}{4 \times 0.0541} = \frac{1.69}{0.2164} = 7.81$$

$$\therefore \quad a = 2.79 \text{ A°}$$

From this value of lattice constant, the interplaner distance d for a particular set of parallel planes having Miller indices (hk$l$) can be obtained using the equation,

$$d_{hkl} = \frac{a}{\sqrt{h^2 + k^2 + l^2}}$$

The interplaner spacing between (100) planes is $d_{100} = \dfrac{2.79°}{\sqrt{1}}$

i.e. $d_{100} = 2.79$ A°

The interplaner spacing between (110) planes is $d_{110} = \dfrac{2.79°}{\sqrt{2}}$,

i.e., $d_{110} = 1.973$ A°

The interplaner spacing between (111) planes is $d_{111} = \dfrac{2.79°}{\sqrt{3}}$,

i.e., $d_{111} = 1.61$ A°

∴ $d_1 : d_2 : d_3 = 1 : \dfrac{1}{\sqrt{2}} : \dfrac{1}{\sqrt{3}}$

The ratio $d_1 : d_2 : d_3$ can be confirmed by Bragg's condition.

For first order, Bragg's condition is $2d \sin \theta = \lambda$.

∴ $\lambda = 2d_1 \sin \theta_1,\ \lambda = 2d_2 \sin \theta_2$ and $\lambda = 2d_3 \sin \theta_3$

$$d_1 : d_2 : d_3 = \dfrac{1}{\sin \theta_1} : \dfrac{1}{\sin \theta_2} : \dfrac{1}{\sin \theta_3}$$

$$= \dfrac{1}{0.3289} : \dfrac{1}{0.4651} : \dfrac{1}{0.5697}$$

$$= 3 : 2.15 : 1.75$$

$$= 1 : 0.716 : 0.58$$

$$= 1 : \dfrac{1}{\sqrt{2}} : \dfrac{1}{\sqrt{3}}$$

The $d_1 : d_2 : d_3$ shows that the given structure is simple cubic crystal.

(B) CHARACTERIZATION TECHNIQUES

Introduction of Characterization Techniques   (Oct. 16, 15)

- For understanding the atomic and molecular structure, the nature of chemical bonding, imperfections, impurity present in the material, etc. one requires a complete knowledge of different types of spectroscopic and techniques for qualitative and quantitative analysis.

- It is our common experience that, when sodium chloride is introduced in the flame of Bunsen burner, a yellow coloured flame is seen. When this flame is examined through spectroscope using grating, two prominent lines having wavelengths 5890 A° and 5896 A° are observed. This shows that, it is possible to identify a material by measuring the wavelengths of light, when the material is introduced in the flame. This is the example of qualitative analysis by emission spectroscopy. By measuring intensities of spectral lines, it is also possible to find the amount of substance producing these lines. This is the example of quantitative analysis by emission spectroscopy.

- In order to understand the properties of material like electric, chemical, thermal, optical of the material, the information about the chemical composition, content of impurity, imperfections etc. of the material must be known. There are number of techniques developed and used by researcher which have found wide applications in the area of characterization of materials.

Some major techniques are given below:

(i)   Conventional chemical methods.

(ii)   Spectrochemical methods: The emission and absorption spectra of all atoms are characteristic and have been investigated. The wavelengths as well as intensities of the emission and absorption spectral lines of different atoms are well tabulated. The major and minor constituents in the sample given for analysis can be detected by the presence of their characteristic emission or absorption spectral lines. Comparing the intensity of the emission line of the sample with that of standard curve, it is possible to find the concentration of impurity present in the sample. The spectrochemical methods include spectrochemical emission analysis, UV-visible absorption spectroscopy, infrared absorption spectroscopy etc.

(iii) Fluorescence analysis method: In this method, radiations used to excite fluorescence could be ultraviolet radiations, gamma rays, accelerated particles like electrons, X-rays. In this method, the photons of electromagnetic radiations are absorbed by molecules, raising them to some excited state. On returning the molecule from excited state to the ground state, it emits radiation. This phenomenon is called **luminescence**. It includes fluorescence and phosphorescence. In fluorescence, the transitions do not involve a change in electron spin whereas in phosphorescence transitions involve change in electron spin. The phosphorescence is slower than fluorescence.

(iv) Mass spectrometry: This method is very sensitive method for trace element analysis. In this method, specimen under investigation is ionized by an ion source. The ions are accelerated by using electric field. The accelerated ions then are allowed to pass through external magnetic field which separates ions of different masses.

(v) X-ray powder diffractometry: An ideal crystal has a perfect periodic arrangement of atoms and molecules. However in practice, actual crystal does not satisfy this requirement. Some atoms or molecules of the material may occupy position that deviate from the ideal arrangement. The methods most commonly used to characterize the given material in regard to their perfection are X-ray diffraction technique, field emission spectroscopy, electron microscopy etc.

(vi) Thermal analysis techniques : There are various methods of thermal analysis. One of the major method is thermogravimetric analysis (TGA). From the analysis of thermal data the details about thermal stability, compositional analysis can be obtained. In addition to above widely used method for analysis and characterization, there are specialized techniques like Nuclear magnetic resonance spectroscopy, Mossbauer spectroscopy, photoelectron spectroscopy, Raman spectroscopy, Scanning electron microscopy, Auger electron spectroscopy etc.

2.9 Thermal Gravimetric Analysis (TGA)   (April 17, 16; Oct. 16)

- Thermogravimetric analysis or thermal gravimetric analysis (TGA) is a method of thermal analysis. In this method changes in physical and chemical properties of materials are measured as a function of increasing temperature (with constant heating rate), or as a function of time (with constant temperature and/or constant mass loss). TGA can provide information about physical phenomena, such as second-order phase transitions, including vaporization, sublimation, absorption, adsorption, and desorption. Likewise, TGA can provide information about chemical phenomena including chemisorptions, desolvation (especially dehydration), decomposition, and solid-gas reactions (e.g. oxidation or reduction).

- Gravimetric analysis is the oldest and accurate method though it is tedious and time consuming. The final step in the method is weighing hence the name given gravimetric analysis. In this method, the constituent to be determined is separated from all other constituents of the given sample in the form of insoluble ppt by using suitable reagent. By the process of filtration and drying, the solvent is also removed leaving only solid ppt of the required constituent of definite chemical composition. The solid ppt is weighed. From the atomic and formula weight, the weight of the constituent is calculated.

  To understand gravimetric analysis consider the following simple example :

- A soluble sodium chloride (NaCl) when treated with soluble silver nitrate (AgNO$_3$), it gives a white ppt of silver chloride (AgCl) according to the equation

$$NaCl \ + \ AgNO_3 \ \rightarrow \ NaNO_3 \ + \ AgCl$$
$$(58.448) + (169.88) \quad (84.999) \ + \ (143.337)$$

  The numbers in the bracket indicate formula weights of substance.

  The above reaction takes place according to weight proportion.

- This equation shows that 143.337 parts of weight of AgCl are always produced by 58.448 parts by weight of sodium chloride (NaCl) when an excess of AgNO$_3$ is added to NaCl solution.

- Suppose a given solution of NaCl when treated with excess AgNO$_3$, solution gives a ppt of AgCl. **When the ppt is washed and dried its mass was found to be 0.2500 gm. Calculate the weight of NaCl in the given solution.**

We know that reaction always take place according to weights proportion.

143.337 gm of AgCl is produced by 58.448 gm of NaCl

$$\therefore \quad 0.2500 \text{ gm of AgCl produces } \frac{0.2500}{143.337} \times 58.448 = 0.1019 \text{ gm of NaCl}$$

- In thermogravimetric analysis (TGA), the temperature of a substance is gradually increased and its weight is noted at different temperatures. From the recorded data, it is possible to know at what temperature, the ppt is decomposing.

- The thermogravimetric analysis can be well understand from the following example :

  When ppt of calcium oxalate is heated to about 475°C, it is found that after this temperature, with increase in temperature, weight of the ppt rapidly decreases. This is because at temperature 455°C, the decomposition of calcium oxalate into $CaCO_3$ starts. Within the temperature range 510°C to 530°C the weight of the sample again remains constant. After the temperature of 530°C, with increase in temperature, the mass of the ppt again decreases. This is because the decomposition of $CaCO_3$ into CaO starts at 530°C. Thus, it is possible to know at what temperature the ppt is decomposing by using TGA technique.

- Thus TGA is a technique whereby the weight of a substance, in an environment heated or cooled at a controlled rate is recorded as a function of time or temperature.

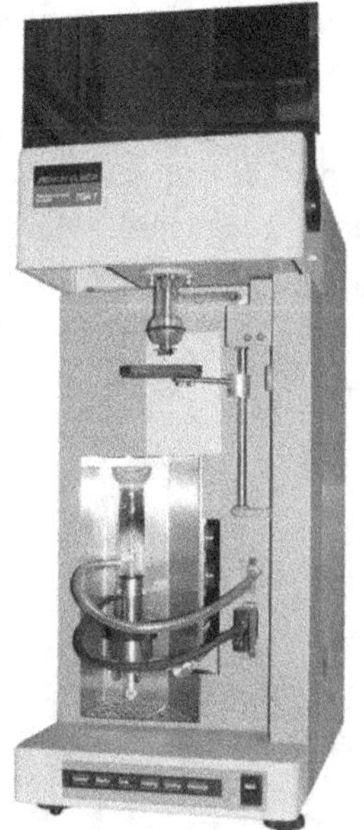

(a) A typical TGA system

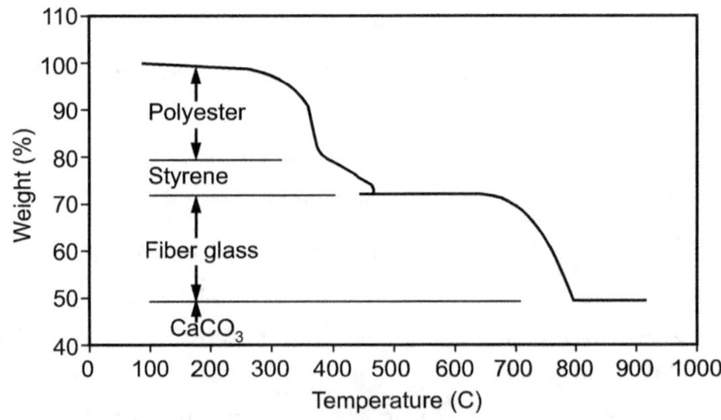

(b) Typical TGA curve of a polymer

Fig. 2.11

- A typical TGA instrument as shown in Fig. 2.11 (a) continuously weighs a sample as it is heated to temperatures upto 2000°C for coupling with FTIR and mass spectrometry gas analysis. As the temperature increases, various components of the sample are decomposed and the weight percentage of each resulting mass change can be measured. Results are plotted with temperature on the X-axis and mass loss on the Y-axis [Refer Fig. 2.11 (b)]. The data can be adjusted using curve smoothing and first derivatives are often also plotted to determine points of inflection for more in-depth interpretations. TGA instruments can be temperature calibrated with melting point standards or Curie point of ferromagnetic materials such as Fe or Ni. A ferromagnetic material is placed in the sample pan which is placed in a magnetic field. The standard is heated and at the Curie point the material becomes paramagnetic which nullifies the apparent weight change effect of the magnetic field.

Uses of TGA analysis :   **(April 16)**

- It can be used to determine temperature and weight change of decomposition reactions, which often allow quantitative composition analysis. It may be used to determine the water content in the sample.

- It allows analysis of reactions with air, oxygen, or other reactive gases.

- It can be used to measure evaporation rates, such as to measure the volatile emissions of liquid mixtures.

- It allows determination of Curie temperatures of magnetic transitions by measuring the temperature at which the force exerted by a nearby magnet disappears on heating or reappears on cooling.

- It is used to measure the weight of fiberglass and inorganic fill materials in plastics, laminates, paints, primers, and composite materials by burning off the polymer resin. The fill material can then be identified by XPS and/or microscopy. The fill material may be carbon black, TiO_2, $CaCO_3$, $MgCO_3$, Al_2O_3, $Al(OH)_3$, $Mg(OH)_2$, talc, Kaolin clay, or silica, for instance.

- It can be used to measure the fill materials added to some foods, such as silica gels and titanium dioxide.

- It can determine the purity of a mineral, inorganic compound, or organic material.

- It distinguishes different mineral compositions from broad mineral types, such as borax, boric acid and silica gels.

2.10 Ultraviolet and Visible Absorption Spectroscopy (April 17, 16)

- For understanding the atomic and molecular structure, the nature of chemical bonding, bond length etc. one requires a thorough knowledge of different types of spectroscopic techniques. **Spectroscopy is the measurement and interpretation of electromagnetic radiations absorbed, scattered or emitted by atoms or molecules or ions.**

- According to quantum mechanics, atoms can exist only in discrete potential energy levels. The potential energy of an atom depends on the electronic configuration. Transition of electrons from one fixed energy level to other fixed energy level leads to emission or absorption of radiation at discrete energies. The frequency at which energy is absorbed or emitted is related with the energy levels involved in the transition and is given by

$$|E_f - E_i| = h\nu$$

- The absorption of visible or ultraviolet radiation by an atomic or molecular species can be expressed by the following equation

$$h\nu + M \rightarrow M^* \qquad\qquad \dots (2.11)$$

  where M^* is an electronically excited species. The life time of M^* (i.e. to remain in the same state) is about 10^{-8} to 10^{-9} sec. Any of several relaxation processes can lead to de-excitation of M^*.

  (i)   The relaxation process involves conversion of excitation energy to heat. In equation form, we can express it as

$$M^* \rightarrow M + heat$$

  (ii)   The relaxation can also occur by decomposition of M^* into new species.

  (iii) The relaxation may involve re-emission of radiation through processes like fluorescence, phosphorescence etc.

$$M^* \rightarrow M + h\nu'$$

- Analytical applications of the absorption of radiation by some chemical species can be either qualitative or quantitative. The qualitative applications of absorption spectrometry depend on the fact that a given molecular species absorbs radiation only in specific regions of the spectrum. The plot of absorption intensity versus wavelength (or frequency) is called an absorption spectrum of given molecular species. This spectrum serves as a finger print for identification of that molecular species.

- The absorption electronic spectra is associated with measurement of energy absorbed when electrons are promoted to higher energy levels.

- The absorption of ultraviolet (UV) and visible radiation by molecules generally occur in one or more electronic absorption bands each of which is made up of many closely packed but discrete lines. Each line in the spectrum arises due to the transition of an electron from ground state to one of the many vibrational energy states associated with each excited electronic state.

- It is found that the excitation of electrons from p and d orbitals occur above 200 nm [(200-380 nm) UV region]. The main difference between UV and visible method is that in UV ions or atoms or molecules absorb more energy for excitation than in the visible region. In short, ultraviolet (UV) spectra are attributed to a process in which outer electrons of atoms absorb radiant energy and undergo transition to higher energy levels. These electronic transitions are quantized and depend on electronic structure of the absorber. The electronic spectra have vibrational and rotational fine structure.

2.11 Ultraviolet Photoelectron Spectroscopy

- In conventional absorption (or emission) spectroscopy, transitions between atomic or molecular levels are studied. The radiation absorbed or emitted in the process is exactly equal to the difference in energy levels. New, let us study photoelectron spectroscopy which mainly includes two techniques, namely ultraviolet photoelectron spectroscopy (UPS) and X-ray photo electron spectroscopy (XPS).

2.11.1 Principle

- Photoelectron spectroscopy is based on photoelectric effect. According to photoelectric effect, the emission of electrons from a material surface is observed when it is illuminated by light or any other radiation of suitable wavelength. The emitted electrons are called as photoelectrons.

- According to energy conservation, if the energy of incident photon is $E = h\nu$, then sum of the ionization energy of the material (or sample) and the kinetic energy of the photoelectron (or ejected electron) is equal to energy of incident photon.

$$\therefore \qquad h\nu = \frac{1}{2} m_e v^2 + I \qquad\qquad \dots (2.12)$$

where, $\frac{1}{2} m_e v^2$ is the kinetic energy of ejected electrons and I is the ionization energy of the sample.

- Using this technique one can measure the ionization energies of molecules, when electrons are ejected from different orbitals of the molecule. The equation (2.12) can be refined in two ways as follows :

(i) The ejected electrons may originate from one of a number of different orbitals and each one has a different ionization energy. Hence, a series of different kinetic energies of the ejected electrons will be obtained. Each photoelectron satisfies the equation given by $h\nu = \frac{1}{2} m_e v^2 + I_i$

where, I_i is the ionization energy of ejected electrons from i^{th} orbital. By measuring K.E. of ejected electrons and knowing frequency of incident photon, the ionization energy can be obtained.

(ii) The ejected electron may leave positive ion in vibrationally excited states. Thus, the incident energy $h\nu$ can be expressed as

$$h\nu = \frac{1}{2} m_e v^2 + I_i + E_{vib}^{+}$$

where, E_{vib} is the energy used to excite the positive ion core into vibration. Each vibrational quantum that is excited leads to a different K.E. of the ejected electron and produces the vibration structure in the photoelectron spectra. The ionization energies of molecules are several electron volts even for valence electrons. So the incident light radiation must have high energy. This requirement is fulfilled by ultraviolet region of the spectrum having wavelength less than 200 nm.

- Therefore, in photoelectron spectroscopy, the incident light used have wavelength always less than 200 nm. When the core electrons of an atom or molecule are to be probed out even high energy is needed and for that purpose X-rays used.

- Ionization that require larger energy leave the emitted electrons with less kinetic energy. On the other hand the ionization that require smaller energy leave the emitted electrons with high kinetic energy. Thus, the pattern of electron kinetic energies are related to the ionization energies of the atoms or molecules of the sample. So in this spectroscopy the kinetic energies of the photoelectrons are measured as a function of frequency.

- In short, in photoelectron spectroscopy there are two main techniques, namely ultraviolet photoelectron spectroscopy (UPS) and X-ray photoelectron spectroscopy (XPS). UPS usually probes the valence electrons in atom of a solid and XPS the core orbital electrons. In early days, EPS was mainly used for identifying the chemical state of atoms and molecules, therefore UPS was also known as electron spectroscopy for chemical analysis (ESCA).

2.11.2 Instrumentation

Ultraviolet-visible spectrophotometer   **(Oct. 16)**

- The instrument used in ultraviolet-visible spectroscopy is called a UV/V is spectrophotometer. It measures the intensity of light passing through a sample (I), and compares it to the intensity of light before it passes through the sample (I_o). The ratio $\dfrac{I}{I_o}$ is called the *transmittance*, and is usually expressed as a percentage (%T). The absorbance A, is based on the transmittance :

$$A = - \log \frac{\% \, T}{100\%}$$

- The UV-visible spectrophotometer can also be configured to measure reflectance. In this case, the spectrophotometer measures the intensity of light reflected from a sample (I), and compares it to the intensity of light reflected from a reference material (I_o) (such as a white tile). The ratio $\dfrac{I}{I_o}$ is called the reflectance, and is usually expressed as a percentage (%R).

- The basic parts of a spectrophotometer are a light source, a holder for the sample, a diffraction grating in a monochromator or a prism to separate the different wavelengths of light, and a detector. The radiation source is often a Tungsten filament (300-2500 nm), a deuterium arc lamp, which is continuous over the ultraviolet region (190-400 nm), xenon arc lamp, which is continuous from 160-2,000 nm; or more recently, light emitting diodes (LED) for visible wavelengths.

- The detector is typically a photomultiplier tube, a photodiode, a photodiode array or a charge-coupled device (CCD). Single photodiode detectors and photomultiplier tubes are used with scanning monochromators, which filter the light so that only light of a single wavelength reaches the detector at one time. The scanning monochromator moves the diffraction grating to "step-through" each wavelength so that its intensity may be measured as a function of wavelength. Fixed monochromators are used with CCDs and photodiode arrays. As both of these devices consist of many detectors grouped into one or two dimensional arrays, they are able to collect light of different wavelengths on different pixels or groups of pixels simultaneously.

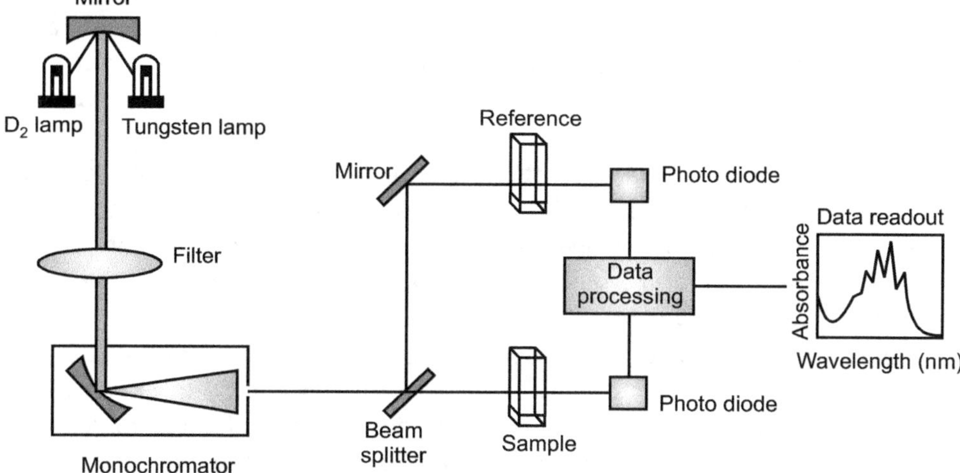

Fig. 2.12 : Schematic of UV- visible spectrophotometer

- A spectrophotometer can be either single beam or double beam. In a single beam instrument if the light passes through the sample cell, I_o must be measured by removing the sample. This was the earliest design and is still in common use in both teaching and industrial labs.

- In a double-beam instrument, the light is split into two beams before it reaches the sample. One beam is used as the reference; the other beam passes through the sample. The reference beam intensity is taken as 100% Transmission (or 0 Absorbance), and the measurement displayed is the ratio of the two beam intensities. Some double-beam instruments have two detectors (photodiodes), and the sample and reference beam are measured at the same time. In other instruments, the two beams pass through a beam chopper, which blocks one beam at a time. The detector alternates between measuring the sample beam and the reference beam in synchronism with the chopper. There may also be one or more dark intervals in the chopper cycle. In this case, the measured beam intensities may be corrected by subtracting the intensity measured in the dark interval before the ratio is taken.

- Samples for UV/Vis spectrophotometry are most often liquids, although the absorbance of gases and even of solids can also be measured. Samples are typically placed in a transparent cell, known as a cuvette. Cuvettes are typically rectangular in shape, commonly with an internal width of 1 cm. (This width becomes the path length, L, in the Beer-Lambert's law.) Test tubes can also be used as cuvettes in some instruments. The type of sample container used must allow radiation to pass over the spectral region of interest. The most widely applicable cuvettes are made of high quality fused silica or quartz glass because these are transparent throughout the UV, visible and near infrared regions. Glass and plastic cuvettes are also common, although glass and most plastics absorb in the UV, which limits their usefulness to visible wavelengths.

- Specialized instruments have also been made. These include attaching spectrophotometers to telescopes to measure the spectra of astronomical features. UV-visible microspectrophotometers consist of a UV-visible microscope integrated with a UV-visible spectrophotometer.

- A complete spectrum of the absorption at all wavelengths of interest can often be produced directly by a more sophisticated spectrophotometer. In simpler instruments the absorption is determined one wavelength at a time and then compiled into a spectrum by the operator. By removing the concentration dependence, the extinction coefficient (ε) can be determined as a function of wavelength.

1. Detection of Impurities : UV absorption spectroscopy is one of the best methods for determination of impurities in organic molecules. Additional peaks can be observed due to impurities in the sample and it can be compared with that of standard raw material. By also measuring the absorbance at specific wavelength, the impurities can be detected. Benzene appears as a common impurity in cyclohexane. Its presence can be easily detected by its absorption at 255 nm.

2. Structure elucidation of organic compounds : UV spectroscopy is useful in the structure elucidation of organic molecules, the presence or absence of unsaturation, the presence of hetero atoms. From the location of peaks and combination of peaks, it can be concluded that whether the compound is saturated or unsaturated, hetero atoms are present or not etc.

3. Quantitative analysis : UV absorption spectroscopy can be used for the quantitative determination of compounds that absorb UV radiation. This determination is based on Beer's law which is as follows :

$$A = \log I_o/I_t = \log 1/T = -\log T = abC = \varepsilon bC$$

where ε is extinction coefficient, C is concentration, and b is the length of the cell that is used in UV spectrophotometer.

4. Qualitative analysis : UV absorption spectroscopy can characterize those types of compounds which absorb UV radiation. Identification is done by comparing the absorption spectrum with the spectra of known compounds.

2.12 Scanning Electron Microscope (SEM)   (Oct. 17, April 17)

- **A scanning electron microscope** is a type of electron microscope that produces images of a sample by scanning it with a focused beam of electrons. The electrons interact with atoms in the sample, producing various signals that can be detected and that contain information about the sample's surface topography and composition. The electron beam is generally scanned in a raster scan pattern, and the beam's position is combined with the detected signal to produce an image. SEM can achieve resolution better than 1 nanometer. Specimens can be observed in high vacuum, in low vacuum, in wet conditions (in environmental SEM), and at a wide range of cryogenic or elevated temperatures.

- The most common SEM mode is detection of secondary electrons emitted by atoms excited by the electron beam. The number of secondary electrons depends on the angle at which beam meets the surface of specimen, i.e. on specimen topography. By scanning the sample and collecting the secondary electrons with a special detector, an image displaying the topography of the surface is created.

- The types of signal produced by scanning electron microscope include characteristic X-rays, light, secondary electrons, back-scattered electrons, transmitted electrons and specimen current. Almost all SEMs have secondary electron detector. The SEM also has facility to detect the signal due to back-scattered electrons, transmitted electron etc. according to the requirement.

- When the beam of electron is incident on the surface of a sample, it interacts with atoms at or near the surface. The signals resulting from interaction are detected. In the most common detection mode, secondary electron imaging (SEI), the SEM can produce very high resolution images of a sample surface giving details about less than 1 to 5 nm in size. As the width of electron beam is very small, SEM micrographs have a large depth of field yielding a characteristic three-dimensional appearance useful for studying and understanding the surface structure of sample.

- Back-scattered electrons (BSE) are beam of electrons that are reflected from the sample by elastic scattering. BSE are also used in analytical SEM along with spectra made from the characteristic X-rays. As the intensity of BSE signal is strongly related with atomic number (Z) of the element in specimen, the BSE images can provide information about the distribution of different elements in the given sample. When energy of electron beam removes an electron from inner shell of the atom of the sample, characteristic X-rays are emitted. These characteristic X-rays are used to identify the composition and measure the abundance of elements in the sample.

- In a typical SEM, an electron beam is thermionically emitted from an electron gun fitted with a tungsten filament cathode. Tungsten is normally used in thermionic electron guns because it has the highest melting point and lowest vapour pressure of all metals, thereby allowing it to be heated for electron emission. It is used because of its low cost. Other types of electron emitters include lanthanum hexaboride (LaB_6) cathodes, which can be used in a standard tungsten filament SEM if the vacuum system is upgraded.

- The electron beam, which typically has an energy ranging from 0.2 keV to 40 keV, is focused by one or two condenser lenses to a spot about 0.4 nm to 5 nm in diameter as shown in Fig. 2.13. The beam passes through pairs of scanning coils or pairs of deflector plates in the electron column, typically in the final lens, which deflect the beam in the x and y axes so that it scans in a raster fashion over a rectangular area of the sample surface.

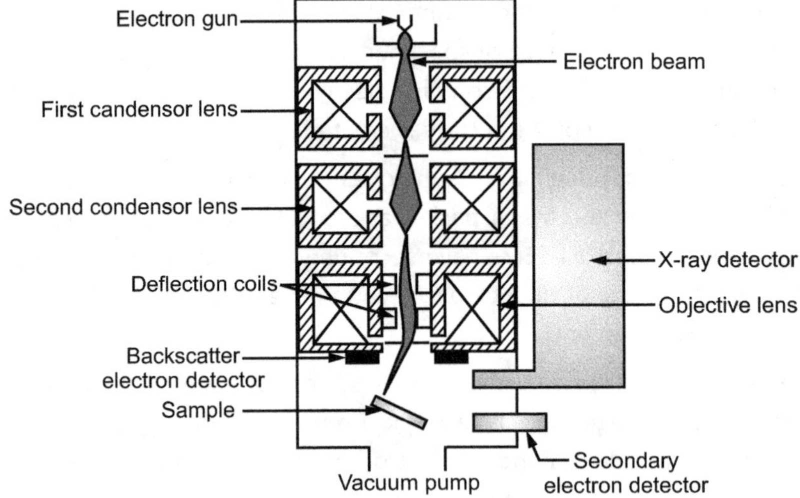

Fig. 2.13 : Schematic of an SEM

- When the primary electron beam interacts with the sample, the electrons lose energy by repeated random scattering and absorption within a teardrop-shaped volume of the specimen known as the interaction volume. It extends from less than 100 nm to approximately 5 μm into the surface.

- The size of the interaction volume depends on the electron's landing energy, the atomic number of the specimen and the specimen's density. The energy exchange between the electron beam and the sample results in the reflection of high-energy electrons by elastic scattering, emission of secondary electrons by inelastic scattering and the emission of electromagnetic radiation, each of which can be detected by specialized detectors. The beam current absorbed by the specimen can also be detected and used to create images of the distribution of specimen current.

- Electronic amplifiers of various types are used to amplify the signals. These signals are displayed as variations in brightness on a computer monitor (or, for vintage models, on a cathode ray tube). Each pixel of computer videomemory is synchronized with the position of the beam on the specimen in the microscope. The resulting image is therefore a distribution map of the intensity of the signal being emitted from the scanned area of the specimen. In older microscopes, image may be captured by photography from a high-resolution cathode ray tube, but in modern machines image is saved to a computer data storage.

Applications of SEM :   **(April 17)**

- The SEM is routinely used to generate high-resolution images of shapes of objects and to show spatial variations in chemical compositions :

    1. Acquiring elemental maps or spot chemical analyses.

    2. Discrimination of phases based on mean atomic number (commonly related to relative density), and

    3. Compositional maps based on differences in trace element "activitors" (typically transition metal and rare earth elements).

- The SEM is also widely used to identify phases based on qualitative chemical analysis and/or crystalline structure. Precise measurement of very small features and objects down to 50 nm in size is also accomplished using the SEM. Backscattered electron images (BSE) can be used for rapid discrimination of phases in multiphase samples. SEMs equipped with diffracted backscattered electron detectors (EBSD) can be used to examine microfabric and crystallographic orientation in many materials.

2.12.1 Scanning Tunneling Electron Microscope   (Oct. 16)

- A new type of microscope called scanning tunneling microscope (STM) has been first built by G. Binning and H. Rohrer in 1983 for monitoring the surface structure of materials.

- **In this microscope, the specimen whose surface is to be looked into is made in the form of one electrode and a metallic material with sharp needle is used as other electrode as shown in Fig. 2.14.** The two electrodes are brought very close to each other and a potential is applied between the two electrodes so that a tunneling current flows.

- In Fig. 2.14, the sharp tip of the needle is brought very close towards the surface which is to be scanned in such a way that electron clouds of the needle and the surface gently touch each other (Refer Fig. 2.14). A voltage applied between the two electrons causes electrons to flow through the narrow channel in the electron clouds. The flow of electrons between the two electrodes is called tunneling current.

- As the density of electrons in the electron cloud falls exponentially with distance, the tunneling current is very sensitive to the distance between the tip of the needle and surface of the specimen. A change in distance of the order of atomic diameter of the single atom causes the tunneling current by a factor of 1000.

- The tip of the needle is scanned along the counters of the surface of the specimen. As the tip is swept across the surface, computerized controlled feedback mechanism senses the tunneling current and maintains the tip of the electrode at constant height above the surface atoms. Thus, the tip follows the countours of the surface of the specimen. The motion of the tip of the electrode is read and processed by a computer and displayed on the screen.

- Thus, a three-dimensional image of the surface can be obtained by sweeping the tip through a pattern of parallel lines separated by small distance (shown in Fig. 2.14) on the surface of the specimen.

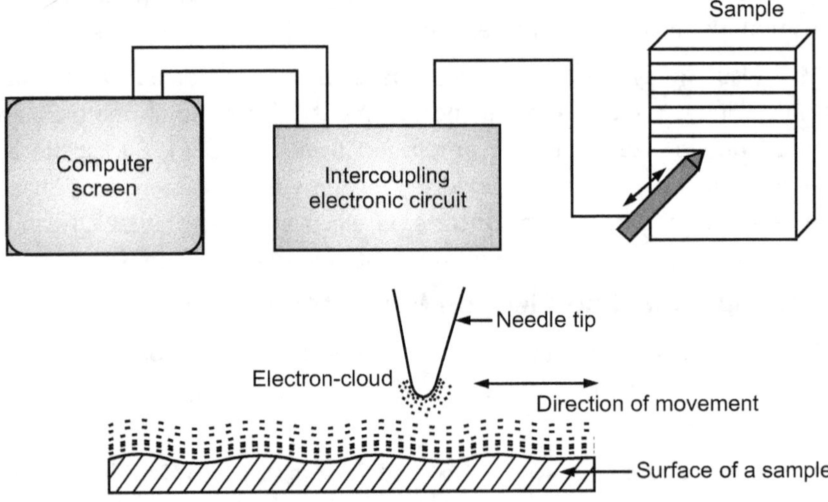

Fig. 2.14 : Scanning tunneling electron microscope

- The scanning tunneling microscope (STM) is used to reveal the surface structure of materials.

Solved Examples

Example 2.1 : *A monochromatic beam of X-ray having wavelength 1.2 A° is incident on a crystal. The first order maxima of reflected X-rays are obtained at an angle 7°. Calculate the glancing angle for second order reflection.*   **(Oct. 15)**

Solution : Bragg's diffraction condition is $2d \sin \theta = n\lambda$.

For first order $2d \sin \theta_1 = \lambda$ and for second order maxima $d \sin \theta_2 = \lambda$.

$$\frac{\sin \theta_1}{\sin \theta_2} = \frac{1}{2}$$

or

$$\sin \theta_2 = 2 \sin \theta_1 = 2 \sin (7°)$$

$$= 2 \times 0.1219 = 0.2438$$

$\therefore$ $$\theta_2 = \sin^{-1} 0.2438 = \mathbf{14.11°}$$ ... **Ans.**

Example 2.2 : *In case of certain crystal the maxima of reflected X-rays are obtained at glancing angles 5° 12', 7° 24' and 9° 1' respectively from three different reflecting planes. Determine the type of cubic crystal.* **(Oct. 16)**

Solution : For first order maxima, Bragg's diffraction condition is

$$2d \sin \theta = \lambda$$

or

$$d = \frac{\lambda}{2 \sin \theta}$$

But λ is constant during the experiment,

$\therefore$

$$d \propto \frac{1}{\sin \theta}$$

$\therefore$ $\quad d_1 \propto \dfrac{1}{\sin \theta_1},\ d_2 \propto \dfrac{1}{\sin \theta_2},\ \text{and}\ d_3 \propto \dfrac{1}{\sin \theta_3}$

or

$$d_1 : d_2 : d_3 = \frac{1}{\sin \theta_1} : \frac{1}{\sin \theta_2} : \frac{1}{\sin \theta_3}$$

$$= \frac{1}{\sin (5° 12')} : \frac{1}{\sin (7° 24')} : \frac{1}{\sin (9° 1')}$$

$$= \frac{1}{0.0906} : \frac{1}{0.1288} : \frac{1}{0.1567}$$

$$= 11.038 : 7.764 : 6.382$$

$$= 1 : 0.703 : 0.578$$

$$= 1 : \frac{1}{\sqrt{2}} : \frac{1}{\sqrt{3}}$$

The ratio $d_1 : d_2 : d_3 = 1 : \dfrac{1}{\sqrt{2}} : \dfrac{1}{\sqrt{3}}$ shows that the crystal is a simple cubic crystal. ... **Ans.**

Example 2.3 : *The distance between (111) planes in a face centred cubic crystal is 2 A°. Determine the lattice parameter and atomic diameter.* **(April 16)**

Solution : The relation between the distance between planes and lattice parameter is

$$d_{hkl} = \frac{a}{\sqrt{h^2 + k^2 + l^2}}$$

$\therefore$
$$d_{111} = \frac{a}{\sqrt{(1)^2 + (1)^2 + (1)^2}}$$

Given,
$$d_{111} = 2 \ A°$$

$\therefore$
$$a = \sqrt{3} \ d$$

$\therefore$
$$a = 2\sqrt{3} \ A°$$

For FCC structure,
$$r = \frac{a}{2\sqrt{2}}$$
where r - radius of atom

The atomic diameter is,

$$D = 2r = \frac{a}{\sqrt{2}}$$

$\therefore$
$$D = \frac{2\sqrt{3}}{\sqrt{2}}$$

$\therefore$
$$D = \sqrt{6} = \textbf{2.45 A°}$$
... Ans.

Example 2.4 : *A BCC crystal is used to measure the wavelength of some X-rays. The Bragg angle for first order reflection from (110) planes is 20.2°. What is the wavelength ? The lattice parameter of the crystal is 3.15 A°.* **(Oct. 15)**

Solution : The distance between (110) planes is

$$d_{hkl} = \frac{a}{\sqrt{h^2 + k^2 + l^2}}$$

Given,
$$a = 3.15 \ A°$$

$\therefore$
$$d_{110} = \frac{3.15}{\sqrt{(1)^2 + (1)^2 + (0)^2}}$$

$$= \frac{3.15}{\sqrt{2}}$$

$\therefore$
$$d_{110} = 2.227 \ A°$$

From Bragg's law, $2d \sin \theta = n\lambda$,   Here n = 1

$\therefore$
$$\lambda = 2d \sin \theta$$

$$= 2 \times 2.227 \times \sin 20.2$$

$\therefore$
$$\lambda = \textbf{1.54 A°}$$
... Ans.

Example 2.5 : *X-rays with wavelength of 1.54 A° are used to calculate the spacing of (200) planes in aluminium. The Bragg angle for first order reflection is 22.4 A°. What is the size of unit cell of the aluminium crystal ?*   **(Oct. 17)**

Solution : Given : λ = 1.54 A°, θ = 22.4°, n = 1

From Bragg's law,

$$n\lambda = 2d \sin\theta$$

$\therefore$
$$d = \frac{\lambda}{2 \sin\theta} \qquad \text{Here } n = 1$$

$$= \frac{1.54}{2 \sin 22.4°}$$

$\therefore$
$$d = 2.02 \text{ A°}$$

Also,
$$d_{hkl} = \frac{a}{\sqrt{h^2 + k^2 + l^2}}$$

$\therefore$
$$d_{200} = \frac{a}{\sqrt{(2)^2 + 0 + 0}}$$

$\therefore$
$$a = 2d$$

$$= 2 \times 2.02$$

$\therefore$
$$a = 4.04 \text{ A°}$$

$$\text{Volume of unit cell} = a^3$$

$$= (4.04 \times 10^{-10} \text{ m})^3$$

$$= \mathbf{65.94 \times 10^{-30} \text{ m}^3} \qquad \textbf{... Ans.}$$

Example 2.6 : *For certain BCC crystal, the (110) planes have separation of 1.181 A°. These (110) planes are irradiated with X-rays of wavelength 1.54 A°. How many orders of Bragg reflections can be observed ?*

Solution : According to Bragg's law,

$$2d \sin\theta = n\lambda$$

$\therefore$
$$n = \frac{2d \sin\theta}{\lambda}$$

Now, maximum possible value of $\sin\theta$ = 1

$\therefore$
$$n = \frac{2d}{\lambda}$$

$$= \frac{2 \times 1.181}{1.54}$$

$\therefore$
$$n = \mathbf{1.53} \qquad \textbf{... Ans.}$$

As the fraction is meaningless, only first order reflection is possible.

Example 2.7 : *An X-ray analysis of crystal is made with monochromatic X-rays of wavelength 0.6 A°. Bragg's reflections are obtained at an angle of (a) 6.45° (b) 9.15° and (c) 13°. Calculate the interplaner spacing of the crystal.*   **(April 17)**

Solution : From Bragg's law,

$$2d \sin \theta = n\lambda$$

$$\therefore \quad \frac{d}{n} = \frac{\lambda}{2 \sin \theta}$$

(a)
$$\frac{d}{n} = \frac{0.6 \times 10^{-10}}{2 \times \sin 6.45°}$$

$$= \mathbf{2.65 \times 10^{-10} \ m} \qquad \text{... Ans.}$$

(b)
$$\frac{d}{n} = \frac{0.6 \times 10^{-10}}{2 \times \sin 9.15°}$$

$$= \mathbf{1.875 \times 10^{-10} \ m} \qquad \text{... Ans.}$$

(c)
$$\frac{d}{n} = \frac{0.6 \times 10^{-10}}{2 \times \sin 13°}$$

$$= \mathbf{1.32 \times 10^{-10} \ m} \qquad \text{... Ans.}$$

The value of d/n in (a) is twice that in (c), it shows that angles 6.45° and 13° represent the first and second order reflection maxima from one set of parallel planes with a spacing which may be found from (a) by putting n = 1 or from (c) by putting n = 2.

Therefore, consider case (a).

$$\frac{d}{n} = 2.65 \times 10^{-10} \ m$$

Here,
$$n = 1$$

$$\therefore \quad d = \mathbf{2.65 \ A°} \qquad \text{... Ans.}$$

Example 2.8 : *Calculate the distance between two lattice planes which give first order diffraction at an angle of 26.42° with molybdenum X-rays of wavelength 0.75 A°.*   **(April 16)**

Solution : From Bragg's law,

$$2d \sin \theta = n\lambda$$

For n = 1, λ = 0.75 A° and θ = 26.42°

$$\therefore \quad d = \frac{n\lambda}{2 \sin \theta}$$

$$= \frac{1 \times 0.75}{2 \times \sin 26.42°}$$

$$\therefore \quad d = \mathbf{0.843 \ A°} \qquad \text{... Ans.}$$

Summary

1. Laue suggested that a crystal which consists of three dimensional array of regularly spaced atoms can serve the purpose of grating.

2. Bragg's condition for reflection is 2d sin θ = nλ.

3. Bragg's diffraction condition in reciprocal lattice is $2\overline{K} \cdot \overline{G} + G^2 = 0$.

4. Bragg's diffraction condition in reciprocal lattice is exactly equivalent to the Bragg's condition in the direct lattice.

5. Experimental methods of X-ray diffraction are (1) Laue method, (2) Rotating crystal method, (3) Debye-Scherrer method.

6. Debye and Scherrer method provides the details about the crystal structure even if the specimen taken for investigation is in the form of polycrystals.

7. When X-rays are reflected back along it's initial path, the process is called back reflection.

8. Cubic structure can be analyzed by powder method.

9. Major characterization techniques are :

    (a) Spectrochemical methods.

    (b) Fluorescence analysis method.

    (c) Mass spectrometry.

    (d) X-ray powder diffractometry.

    (e) Thermal analysis technique.

10. In TGA method, changes in physical and chemical properties of materials are measured as a function of increasing temperature or as a function of time.

11. Spectroscopy is the measurement and interpretation of electromagnetic radiations absorbed, scattered or emitted by atoms or molecules or ions.

12. Photoelectron spectroscopy is based on photoelectric effect. According to photoelectric effect, the emission of electrons from a material surface is observed when it is illuminated by light or any other radiation of suitable wavelength. The emitted electrons are called as photoelectrons.

13. SEM is a type of electron microscope that produces images of a sample by scanning it with a focused beam of electrons. It contains information about sample's surface topology and composition.

14. In STEM microscope, the specimen whose surface is to be looked into is made in the form of one electrode and a metallic material with sharp needle is used as other electrode.

15. STEM is used to reveal surface structure of materials.

Exercise

(A) Short Answer Type Questions :

1. Explain crystal as a grating.
2. What are X-rays ?
3. State Bragg's condition for diffraction in direct lattice.
4. State Bragg's condition for diffraction in reciprocal lattice.
5. Define reciprocal lattice.
6. Why ordinary optical grating cannot diffract X-rays ?
7. How is the reciprocal lattice constructed in direct lattice ?
7. If the angle between the diffraction of incident X-rays and diffracted one is 20°, what is the angle of incidence ?
8. Explain how crystal can be used as a diffraction grating for X-rays.
9. State various spectroscopic techniques in analysis of crystal structure.
10. What is the principle involved in gravimetric analysis ?
11. What do you mean by qualitative analysis ?
12. What do you mean by quantitative analysis ?
13. Explain the term spectroscopy.
14. What is the main difference between ultraviolet and visible spectroscopy ?
15. Give the principle of photoelectron spectroscopy.
16. What do you mean by STM ? Give application of STM.
17. What is the main difference between UPS and XPS ?
18. What are advantages of powder method ?
19. What do you mean by SEM ? Give application of SEM.

(B) Long Answer Type Questions :

1. Explain Bragg's law associated with diffraction of X-rays by crystals.
2. Derive the Bragg's condition for diffraction.
3. Show that Bragg diffraction condition in the reciprocal lattice

$$2\,\vec{K}\cdot\vec{G} + G^2 = 0$$

4. With the help of Ewald's construction show that the diffraction condition in reciprocal lattice is exactly equivalent to $2d\sin\theta = n\lambda$ in the direct lattice.
5. Explain how X-rays are used for determination of crystal structure.

6. Describe powder method for determination of crystal structure.

7. Describe the X-ray powder diffraction camera and explain how it is used to determine crystal structure.

8. With suitable diagram explain scanning electron microscope.

9. What do you mean by gravimetric analysis ? With suitable example explain gravimetric analysis.

10. What do you mean by TGA ? By taking suitable example explain TGA.

11. State characterization techniques and explain applications of each technique.

12. Write a short note on UV-visible absorption spectroscopy.

13. Give the principle of photoelectron spectroscopy.

14. With suitable diagram, explain UV photoelectron spectrometer.

15. By taking suitable data, give analysis of cubic crystal by powder method.

16. What is SEM ? Explain principle and applications of SEM.

(C) Unsolved Problems :

1. The X-rays of wavelength 1.6 A° are diffracted by a Bragg crystal spectrometer at an angle of 14.2° in the first order. What is the spacing of atomic layers in the crystal ?

    (Ans. 2.15 A°)

2. The X-rays of wavelength 1.392 A° are reflected from face of NaCl crystal. The first order reflection is observed at an angle of 14° 17′ 26″ . Calculate the lattice spacing.

    (Ans. 2.820 A°)

3. The glancing angle for the first order spectrum is 5° 30' in Bragg's X-ray spectrometer. If the crystal spacing d = 2.65 A°, calculate the wavelength of X-rays.

    (Ans. 0.5 A°)

4. A powder pattern is obtained for lead with radiation of X = 1.54 A°. The (220) reflection is observed at Bragg angle θ = 32°. What is the lattice parameter of lead and the radius of atom ?   **(Ans.** a = 4.1 A°, r = 1.45 A°)

5. Calculate largest and smallest Bragg's angles for reflection of X-rays of wavelength 1.436° from simple cubic structure if edge of unit cell is 3.6 A°. Consider reflection from the parallel (100) planes of the crystal.   **(Ans.** θ_S = 11° 24', θ_L = 84° 30')

6. The edge of the cubic unit cell of lead is 4.92 A°. The density of lead is 11.55 g/cm³ and its molecular weight is 207.19.

    (a) What type of cubic lattice does lead exhibit ?

    (b) Calculate the smallest angle for X-rays 0.708 A° reflected from (110) planes of this crystal.   **(Ans.** n = 4, θ = 11° 44')

7. What will be the wavelength of X-ray which gives a diffraction angle 2θ equal to 16.8° for a crystal, if minimum interplaner distance in the crystal is 0.400 nm and only second order diffraction is observed ? **(Ans.** 0.584 A°)

8. At what angle would a first order reflection be observed in the X-ray diffraction of a set of crystal planes for which d = 0.303 nm, if X-rays used have wavelength of 0.071 nm. At what angle would second order reflection from the set of plane is observed ? **(Ans.** $\theta_{first\ order}$ = 6.73°, $\theta_{second\ order}$ = 13.55°)

9. The distance between adjacent atomic planes in calcite is 0.3 nm. What is the smallest angle between these plane and an incident beam of 30 pm X-ray at which scattered X-ray can be detected. **(Ans.** 8.6°)

10. Calculate longest wavelength that can be analyzed by rock salt of crystal of spacing 2 A° in the first order and second order.

    (Ans. For n = 1, λ = 4 A° and for n = 2, λ = 2 A°)

11. A beam of 0.154 nm X-rays are directed at certain planes of a silicon crystal. As angle of incidence increases from zero, the first strong interference occurs when a beam makes an angle of 34.5° with the planes. (a) How far apart are the planes ? (b) Will you find other interference maxima from these planes from greater angle of incidence?

    (Ans. (a) 0.136 nm, (b) value of n of 2 or greater gives sin θ than unity which is impossible. Hence there is no other angle for interference maxima for this particular set.)

12. Calculate distance between two lattice planes which give first order diffraction at an angle of 25° with molybdenum X ray of wavelength 0.7 A°. **(Ans.** 0.828 A°)

Chapter **3**...

Free Electron and Band Theory of Metals

Contents ...

Edwin Herbert Hall (1855-1938)

The **Hall Effect** was discovered by Edwin Hall in 1879, while working on his doctoral thesis in Physics. Hall's experiments consisted of exposing thin gold leaf (and, later, using various other materials) on a glass plate and tapping off the gold leaf at points down its length.

The effect is a potential difference (Hall voltage) on opposite sides of a thin sheet of conducting or semiconducting material (the Hall element) through which an electric current is flowing. This was created by a magnetic field applied perpendicular to the Hall element. The ratio of the voltage created to the amount of current is known as the *Hall resistance*, and is a characteristic of the material in the element.

3.1 Introduction

- Free electron theory of metals has been developed in following three stages :

  (i)   The classical free electron theory of Drude and Lorentz in 1900, postulated that the metals consist of positive ion core with valence electrons moving freely inside the metals. The behaviour of free electrons moving inside the metals is similar to that of gas molecules in a vessel. The free electrons obey the laws of classical Maxwell-Boltzmann statistics.

  (ii) The quantum free electron theory was proposed by Sommerfeld (1928) in which, he treated the problem quantum mechanically using Fermi-Dirac statistics rather than classical-Maxwell Boltzmann statistics.

  (iii) In the zone theory, electrons move in a periodic field provided by positively charged ion cores.

- We shall discuss all above theories in the present chapter. The present chapter also includes Hall effect and energy bands in solid, distinction between metal, semiconductor and insulator on the basis of band theory.

3.2 Classical Free Electron Model of Metals (Oct. 17, 16, 15; April 17, 16)

- Soon after Thomson's discovery of the electron, P. Drude and Lorentz proposed the free electron theory of metals in 1990. They tried to explain some of the properties of metal like electrical and thermal conductivities on the basis of this theory. According to this theory, the physical properties of metal may be explained by modeling the metal as a classical gas of conduction electrons moving through a fixed lattice of ion cores. The classical free electron theory is based on following assumptions:

  1.   A metal contains a number of free electrons.

  2.   Most weakly bound i.e. valence electrons of constituent atoms move freely throughout the volume of the metal like gas molecules in a vessel.

  3.   The mutual repulsion between the electrons is ignored.

  4.   The potential field due to positive ions is assumed to be completely uniform so that electrons can move from place to place in the solid without any change in the energy.

  5.   They collide occasionally with atoms and at any given temperature, their velocities could be determined according to Maxwell-Boltzmann distribution law.

- In particular, the model predicts the correct functional form of Ohm's law and the empirical connection between the electrical and thermal conductivities of metal, known as Wiedemann-Franz law. However, the model does not predict accurately the values of electrical conductivity and thermal conductivities of most of the metals that are observed experimentally.

Electrical Conductivity :

- Ohm's law is given by $J = \sigma E$

  where J is current density, E is the electric field and σ is the electrical conductivity.

- We can show that Ohm's law follows from considering a conductor consisting a classical gas of free electrons moving through a fixed lattice of positive ions. The electrons in a metal move randomly along straight line path between two successive collisions of an electron with lattice ions (Refer Fig. 3.1).

- The root mean square (r.m.s.) velocity is fairly high at room temperature and may be calculated from classical equipartition theorem. According to this theorem, we have

$$\frac{1}{2}\,m\overline{v^2} = \frac{3}{2}\,k_B T$$

where $\overline{v^2}$ is mean square velocity of the electron, T is the temperature and k_B is Boltzmann constant.

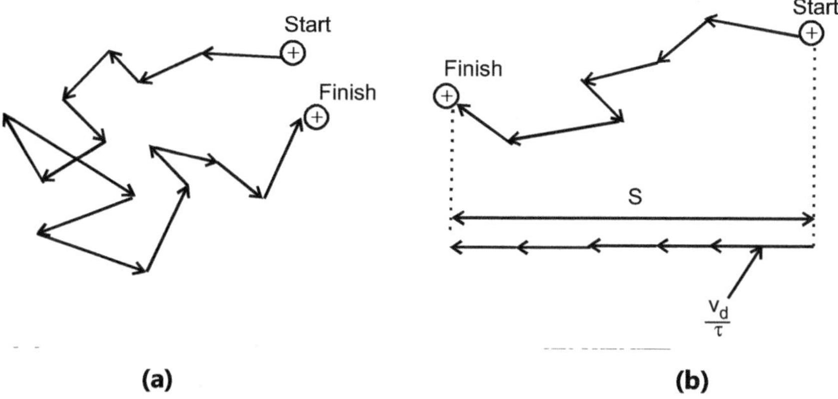

Fig. 3.1 : Movement of free electron through a fixed lattice

- Therefore, the r.m.s. velocity is given as

$$v_{rms} = \sqrt{\overline{v^2}}$$

or
$$v_{rms} = \left(\frac{3k_B T}{m}\right)^{1/2} \qquad \text{... (3.1)}$$

- The path between the collisions shown in Fig. 3.1 (a) is free path and **the average distance travelled by the electron between two collisions is called mean free path.** The mean free path L is related to mean free time τ, the average time between two collisions by the equation

$$L = v_{rms}\,\tau \qquad \text{... (3.2)}$$

- When an electric field is applied to the material, an electric force is exerted on each electron along the field direction. The magnitude of acceleration produced due to electric field along the field direction is given by

$$a = \frac{eE}{m} \qquad \text{... (3.3)}$$

- Due to electric force exerted on an electron during each step, resulting displacement is small as compared to the mean free path. The cumulative effect of these displacements may be viewed in terms of a small average drift speed superposed on the rather high random velocities as shown in Fig. 3.1 (b).

- The total displacement of the electron after p collisions can be taken as

$$S = \frac{a}{2}\left(t_1^2 + t_2^2 + t_3^2 + ... + t_p^2\right) \qquad \text{... (3.4)}$$

where t_1, t_2, t_3, ..., t_p are the successive times between collisions. It should be noted that we have ignored the random initial velocities in the equation (3.4).

We may write equation (3.4) in terms of averages as

$$S = \frac{a}{2}\, p\, \overline{t^2}$$

It is observed that the average value of t^2 is equal to $2\tau^2$. Therefore, we can write,

$$S = ap\tau^2$$

or $$S = \frac{eEp\tau^2}{m} \qquad \text{... (3.5)}$$

The drift velocity is given by

$$S = pv_d\, \tau$$

$$\therefore \quad v_d = \frac{eE\tau}{m} \qquad \text{... (3.6)}$$

The relation between current density and drift velocity is given by

$$J = nev_d$$

where n is the number of electrons per unit volume all moving with drift velocity.

$$\therefore \quad J = \frac{ne^2\, \tau E}{m}$$

Since $J = \sigma E$, from above equation, we can write the conductivity as

$$\sigma = \frac{ne^2\, \tau}{m}$$

or $$\sigma = \frac{ne^2\, L}{mv_{rms}} \qquad \text{... (3.7)}$$

Substituting equation (3.1) in equation (3.7), we get

$$\sigma = \frac{ne^2 L}{(3k_BTm)^{1/2}} \qquad \dots (3.8)$$

The reciprocal of conductivity is called resistivity and is given by

$$\rho = \frac{(3k_BTm)^{1/2}}{ne^2 L} \qquad \dots (3.9)$$

Thermal Conductivity :

- The rate of flow of heat across the metal bar of cross-sectional area A is given by

$$\frac{dQ}{dt} = -KA\frac{dT}{dx}$$

where dT/dx is the temperature gradient and K is the constant called thermal conductivity.

- Since good electrical conductors are also good thermal conductors, it is usual to assume that the highly mobile classical electron gas transports charge as well as heat energy through metal via a random collision process.

According to the kinetic theory of gases, the thermal conductivity is given by

$$K = \frac{1}{3} C_V v_{rms} L \qquad \dots (3.10)$$

where C_V is specific heat at constant volume and L is the mean free path.

- As we have assumed that the electrons behave like classical gas and obey Maxwell-Boltzmann distribution law, the heat capacity per mole is given by 3R/2 or $3Nk_B/2$, where N is Avogadro's number and k_B is Boltzmann constant. The heat capacity per unit volume can be obtained by dividing the term $\dfrac{3Nk_Bn}{2}$ by N.

$$\therefore \qquad C_V = \left(\frac{3Nk}{2}\right)\left(\frac{n}{N}\right) = \frac{3}{2} k_B n$$

$$\therefore \qquad K = \frac{1}{2} k_B n v_{rms} L$$

Substituting equation (3.1) in above equation, we get

$$K = \frac{1}{2} k_B nL \left(\frac{3kT}{m}\right)^{1/2} \qquad \dots (3.11)$$

Taking ratio of equations (3.11) and (3.8), we get

$$\frac{K}{\sigma} = \frac{3}{2} \frac{k_B^2}{e^2} T \qquad \dots (3.12)$$

- **This equation is called Wiedemann-Franz law.** Equation (3.12) shows that the ratio K/σ is proportional to the temperature T. **The ratio $K/\sigma T$ is a constant and known as the Lorentz number.** According to the classical theory, it has the same value for any metal at any temperature.

- The ratio of $K/\sigma T$ is nearly constant from metal to metal and with varying temperatures but the value $3k^2/2e^2$ does not agree precisely with the measured Lorentz number.

3.3 Sommerfeld's Free Electron Model   (April 16)

- The quantum free electron theory (1928) of Sommerfeld is based on three basic assumptions :

   (i)   Valence electrons of constituent atoms move freely throughout the volume of the metal like gas molecules in a vessel. But these free electrons obey the laws of quantum mechanics rather than classical mechanics. So this model has replaced classical statistics of Maxwell-Boltzmann by Fermi-Dirac statistics in which metal is replaced by a box containing non-interacting free electrons.

   (ii)   The free electrons obey Pauli's exclusion principle.

   (iii)   The free electrons in a solid move as if it is a system of waves.

- Thus, Sommerfeld has assumed that valence electrons travel in a constant potential inside the metal and these electrons are prevented from leaving the solid by very high barriers at the boundaries of crystal. In short, the electrons are trapped in a constant potential well as shown in Fig. 3.2.

- The number of permissible energy states for an electron to occupy in the potential box are calculated by Schrödinger's wave equation. The distribution of free electrons in the various permissible energy states is answered by applying Fermi-Dirac distribution.

3.4 Energy Levels & Density of States in One dimension
(Oct. 17, 16)

- Consider a free electron of mass m confined in a box of length L. Assume the potential energy everywhere within the crystal to be zero. At the two ends of the crystal, the electron is prevented from leaving the crystal by a very high potential energy barrier.

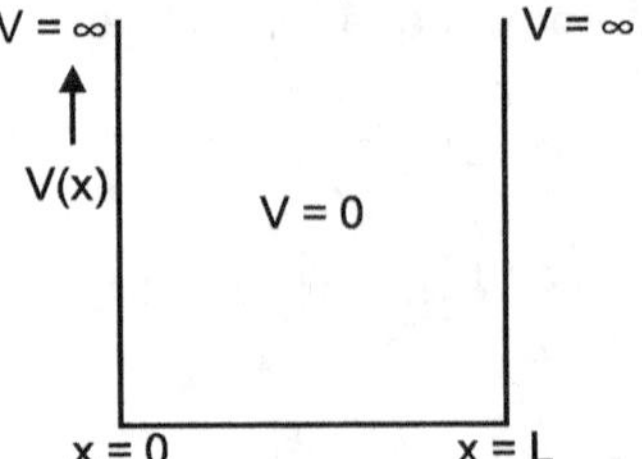

Fig. 3.2 : One-dimensional potential box

- The Schrödinger's time independent equation in one dimension for the region is given by

$$-\frac{\hbar^2}{2m}\frac{d^2\psi}{dx^2} + V\psi = E\psi$$

For region $V = 0$, the equation becomes

$$-\frac{\hbar^2}{2m}\frac{d^2\psi}{dx^2} = E\psi \qquad \ldots (3.13)$$

$$\frac{d^2\psi}{dx^2} + \frac{2mE}{\hbar^2}\psi = 0$$

$$\frac{d^2\psi}{dx^2} + k^2\psi = 0 \qquad \ldots (3.14)$$

where, $$k^2 = \frac{2mE}{\hbar^2} \qquad \ldots (3.15)$$

$$\therefore \qquad k = \sqrt{\frac{2mE}{\hbar^2}}$$

- The solution of equation (3.14) is

$$\psi(x) = A\sin kx + B\cos kx \qquad \ldots (3.16)$$

where A and B are arbitrary constants to be determined by applying boundary conditions. The electron wave function has to satisfy the following boundary conditions:

1. $\psi = 0$ at $x = 0$ and

2. $\psi = 0$ at $x = L$

Applying the first condition, equation (3.16) becomes

$$0 = A\sin k(0) + B\cos k(0)$$

$$0 = 0 + B$$

$$\therefore \qquad B = 0 \qquad \ldots (3.17)$$

The solution given by equation (3.16) becomes,

$$\psi(x) = A\sin kx \qquad \ldots (3.18)$$

Applying the second condition (i.e. $\psi = 0$ at $x = L$),

$$0 = A\sin kL$$

$$\therefore \qquad kL = n\pi$$

or $$k = \frac{n\pi}{L} \qquad \ldots (3.19)$$

where $n = 1, 2, 3, \ldots$ represents the order of the state. $n = 0$ is not allowed.

Therefore, the solution to the Schrödinger's equation in region $0 < x < L$ is

$$\psi_n (x) = A \sin \frac{n\pi}{L} x \qquad \text{... (3.20)}$$

From equations (3.15) and (3.19),

$$\frac{2mE_n}{\hbar^2} = \left(\frac{n\pi}{L}\right)^2$$

$$\therefore \qquad \boxed{E_n = \frac{n^2\pi^2 \hbar^2}{2mL^2}} \qquad \text{... (3.21)}$$

It indicates that, the energy is a function of a quantum number n.

- Fig. 3.3 indicates energy levels and wave functions of a free electron of mass m confined to a line of length L The energy levels are labelled according to quantum number n which gives number of half wavelengths in the wave function.

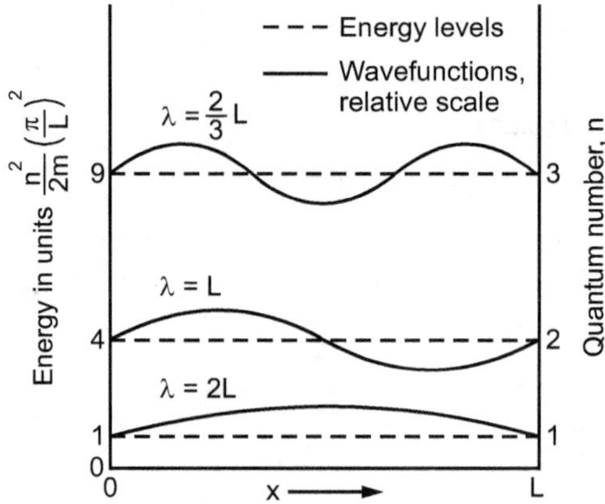

Fig. 3.3: First three energy levels

- With the help of Pauli's exclusion principle, N electrons can be accommodated on the line of length L. According to the **Pauli exclusion principle,** no two electrons can have all their quantum numbers identical. That is, each orbital can be occupied by at most one electron. This applies to electrons in atoms, molecules, or solid.

- Each energy level with quantum number n can have two electrons with opposite spins. Let n_F be the topmost filled energy level.

$$\therefore \qquad N = 2n_F \qquad \text{... (3.22)}$$

- The **Fermi energy** E_{F_0} is defined as the energy of the topmost filled level in the ground state of the N electron system. Thus, Fermi energy level at 0°K,

$$E_{F_0} = \frac{\hbar^2}{2m}\left(\frac{\pi}{L}\right)^2 n_F^2$$

$$= \frac{\hbar^2}{2m}\left(\frac{\pi}{L}\cdot\frac{N}{2}\right)^2 \qquad \text{... (3.23)}$$

- It indicates that, the energy of the topmost level depends upon the number of electrons in the metal and size of the potential box.

3.4.1 Density of States in One Dimension

From equation (3.21),

$$E_n = \frac{\hbar^2}{2m}\left(\frac{n\pi}{L}\right)^2$$

$$\therefore \qquad n^2 = \frac{2mL^2}{\pi^2\hbar^2}\, E_n$$

$$\text{and} \qquad n = \frac{L}{\pi\hbar}\,(2mE_n)^{1/2} \qquad \text{... (3.24)}$$

Differentiating and multiplying above equation by $\frac{1}{2}$, we get the number of quantum states of electrons in the energy range between E and E + dE.

$$dn = \frac{L}{2\pi\hbar}\,(2m)^{1/2}\cdot\frac{1}{2}\,E_n^{-1/2}\,dE$$

$$\therefore \qquad dn = \frac{L}{4\pi}\left(\frac{2m}{\hbar^2}\right)^{1/2} E_n^{-1/2}\,dE \qquad \text{... (3.25)}$$

Let us denote number of quantum states of electrons between E and E + dE by Z(E) dE.

$$\therefore \qquad Z(E)\,dE = dn = \frac{L}{4\pi}\left(\frac{2m}{\hbar^2}\right)^{1/2} E^{-1/2}\,dE$$

$$\therefore \qquad Z(E) = \frac{L}{4\pi}\left(\frac{2m}{\hbar^2}\right)^{1/2} E^{-1/2}$$

Due to electrons having opposite spins, each orbital can be split up into two.

$$\text{Hence} \qquad Z(E) = 2\times\frac{L}{4\pi}\left(\frac{2m}{\hbar^2}\right)^{1/2} E^{-1/2} \qquad \text{... (3.26)}$$

The equation (3.26) represents density of states in one dimension.

3.4.2 Three Dimensional Case

- Consider Schrödinger's three-dimensional time independent equation

$$-\frac{\hbar^2}{2m}\nabla^2\psi + V\psi = E\psi$$

where
$$\psi = \psi(x, y, z)$$

$$-\frac{\hbar^2}{2m}\left(\frac{\partial^2}{\partial x^2} + \frac{\partial^2}{\partial y^2} + \frac{\partial^2}{\partial z^2} + V\right)\psi = E\psi \qquad \text{... (3.27)}$$

- Let us assume that the electrons are confined to a cube having length of edge L as shown in Fig. 3.4. Since the electrons are free inside the box, we have V = 0 inside the box.

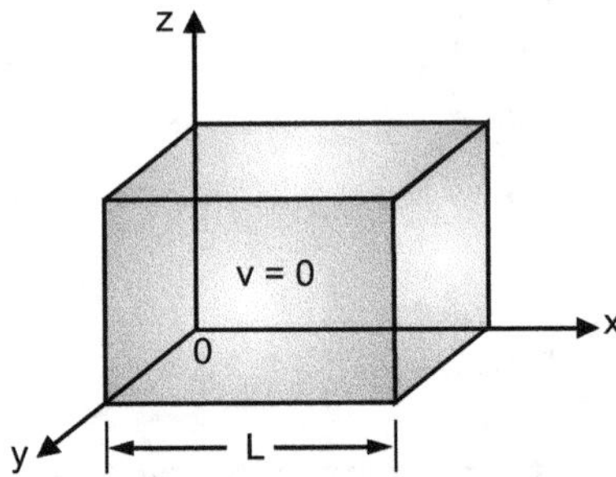

Fig. 3.4 : A three-dimensional potential box

- The Schrödinger's time independent equation for the region where V = 0 is given by

$$-\frac{\hbar^2}{2m}\left(\frac{\partial^2}{\partial x^2} + \frac{\partial^2}{\partial y^2} + \frac{\partial^2}{\partial z^2}\right)\psi = E\psi$$

- Let the wave function ψ be periodic in x, y, z with period L.

Thus,
$$\psi(x + L, y, z) = \psi(x, y, z)$$
$$\psi(x, y + L, z) = \psi(x, y, z)$$
$$\psi(x, y, z + L) = \psi(x, y, z)$$

The boundary conditions are
$$\psi = 0 \text{ at } x = 0, \ y = 0 \text{ and } z = 0$$
$$\psi = 0 \text{ at } x = L, \ y = L \text{ and } z = L$$

- Hence, the solution of the Schrödinger's equation in the region of box is assumed to be of the standing wave type

$$\psi(x, y, z) = A \sin(k_x x)\sin(k_y L)\sin(k_z z)$$

But
$$k_x = \frac{n_x \pi}{L}, \ k_y = \frac{n_y \pi}{L} \text{ and } k_z = \frac{n_z \pi}{L} \qquad \text{(refer equation 3.20)}$$

$$\therefore \qquad \psi(x, y, z) = A\sin\left(\frac{n_x \pi}{L}x\right)\sin\left(\frac{n_y \pi}{L}y\right)\sin\left(\frac{n_z \pi}{L}z\right) \qquad \text{... (3.28)}$$

where n_x, n_y, n_z are three quantum numbers which can take only positive integer values.

The corresponding form of energy is given by (refer equation 3.21)

$$E_{n_x,\, n_y,\, n_z} = \frac{\hbar^2 \pi^2}{2mL^2}(n_x^2 + n_y^2 + n_z^2)$$

$$n_x^2 + n_y^2 + n_z^2 = n^2$$

or

$$k^2 = k_x^2 + k_y^2 + k_z^2$$

$$\therefore \qquad E_n = \frac{\hbar^2 \pi^2}{2mL^2} n^2 \qquad \qquad \dots (3.29)$$

$$\therefore \qquad E_n \propto n^2$$

The energy of an electron inside the box in the nth state (E_n) is directly proportional to square of quantum number (n).

3.4.3 Density of States in Three Dimensions   (April 17)

- In order to calculate the density of electrons in the energy range between E and E + dE, let us draw two concentric spheres of radii n and n + dn in the n-space as shown in Fig. 3.5.

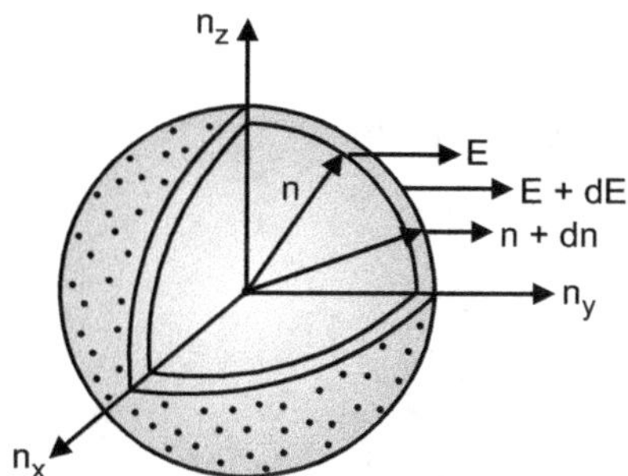

Fig. 3.5 : Spheres representing density of states in n-space

- At any point (n_x, n_y, n_z) with integer values of co-ordinates represent an energy state.

  Thus, all the points on the surface of the sphere of radius n $\left(\text{where } n^2 = n_x^2 + n_y^2 + n_z^2\right)$ will have the same energy.

  Thus, the number of states having energy values between E and E + dE is given by

$$Z(E)\, dE = \frac{1}{8} \times 4\pi n^2\, dn$$

- The factor 1/8 is due to the availability of only one octant of the sphere because of positive integers n_x, n_y, n_z.

$$\therefore \qquad Z(E)\, dE = \frac{\pi}{2} n^2\, dn \qquad \qquad \dots (3.30)$$

But from equation (3.29),

$$E = \frac{\hbar^2 \, n^2 \, \pi^2}{2mL^2}$$

$$\therefore \qquad n^2 = \frac{2mL^2}{\pi^2 \, \hbar^2} \, E \qquad\qquad \text{... (3.31)}$$

$$\therefore \qquad n = (2m)^{1/2} \, \frac{L}{\pi\hbar} \, E^{1/2}$$

$$\therefore \qquad dn = \frac{L}{\pi\hbar} \, (2m)^{1/2} \, \frac{1}{2} \, E^{-1/2} \, dE \qquad\qquad \text{... (3.32)}$$

Substituting equations (3.32) and (3.31) in equation (3.30), we get

$$Z(E) \, dE = \frac{\pi}{2} \, (2m) \, \frac{L^2}{\pi^2 \, \hbar^2} \, E \times \frac{1}{2} \, \frac{L}{\pi\hbar} \, (2m)^{1/2} \, E^{-1/2} \, dE$$

$$= \frac{\pi}{4} \left(\frac{L}{\pi\hbar}\right)^3 (2m)^{3/2} \, E^{1/2} \, dE$$

$$= \frac{\pi L^3}{4\pi^3} \left(\frac{2m}{\hbar^2}\right)^{3/2} E^{1/2} \, dE$$

$$= \frac{L^3}{4\pi^2} \left(\frac{2m}{\hbar^2}\right)^{3/2} E^{1/2} \, dE$$

$$= 2\pi L^3 \left(\frac{2m}{h^2}\right)^{3/2} E^{1/2} \, dE$$

$$= 2\pi V \left(\frac{2m}{h^2}\right)^{3/2} E^{1/2} \, dE$$

where, $V = L^3$ (the volume of the box)

The Pauli's exclusion principle allows two electrons in each state. Hence the numbers of energy levels actually available are

$$Z(E) \, dE = 2 \times 2\pi V \left(\frac{2m}{h^2}\right)^{3/2} E^{1/2} \, dE \qquad\qquad \text{... (3.33)}$$

Thus, density of states, $Z(E) = 4\pi V \left(\dfrac{2m}{h^2}\right)^{3/2} E^{1/2}.$

Z(E) versus E is plotted as shown in Fig. 3.6.

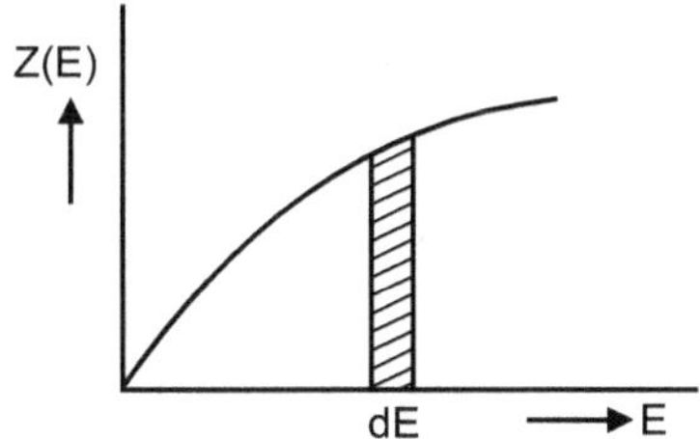

Fig. 3.6 : Density of energy states for a free electron gas

- According to this model, an electron gas behaves like a system of Fermi particle and hence obeys Fermi-Dirac statistics. Sommerfeld considered the distribution of a large number of electrons in thermal equilibrium among the various states in a 3-dimensional box.

The actual number of electrons N(E) dE in a given energy interval dE is given by

$$N(E)\,dE \;=\; Z(E)\,dE\,F(E) \qquad \text{... (3.34)}$$

where F(E) is the Fermi distribution function.

$$F(E) \;=\; \frac{1}{1 + \exp\left(\dfrac{E - E_F}{K_B\,T}\right)}$$

At T = 0, $F(E) = 1$ for all values of $E < E_F$

         $F(E) = 0$ for all values of $E > E_F$

$\therefore$ $$N(E)\,dE \;=\; 4\pi V \left(\frac{2m}{h^2}\right)^{3/2} E^{1/2}\,\frac{dE}{1 + \exp\left(\dfrac{E - E_F}{K_B\,T}\right)} \qquad \text{... (3.35)}$$

The distribution is shown in Fig. 3.7.

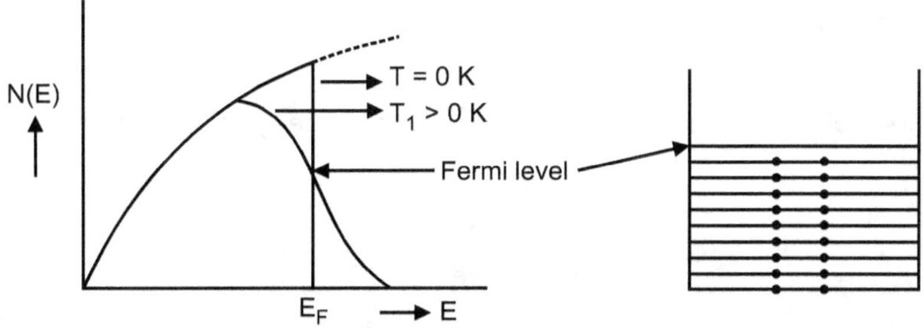

Fig. 3.7 : Density of states as a function of electron energy at different temperatures

The Fermi level is the topmost filled energy level at T = 0.

At T = 0 K, $F(E) = 1$ as $E < E_F$

The total number of electrons upto the Fermi level at T = 0 K is

$$\int N(E)\,dE = \int_0^{E_{F_0}} 4\pi V\left(\frac{2m}{h^2}\right)^{3/2} E^{1/2}\,dE$$

$$N = \frac{2}{3} \times 4\pi V\left(\frac{2m}{h^2}\right)^{3/2} E_{F_0}^{3/2} = \frac{8}{3}\pi V\left(\frac{2m}{h^2}\right)^{3/2} E_{F_0}^{3/2}$$

Hence the Fermi energy at absolute zero is

$$E_{F_0} = \frac{h^2}{2m}\left(\frac{3N}{8\pi V}\right)^{2/3} \qquad \ldots (3.36)$$

At absolute zero, the average energy of an electron is given by

$$\bar{E} = \frac{1}{N}\int_0^{E_{F_0}} E\,N(E)\,dE \qquad \ldots (3.37)$$

$$= \frac{4\pi V\left(\frac{2m}{h^2}\right)^{3/2}}{N}\int_0^{E_{F_0}} E^{3/2}\,dE = \frac{4\pi V\left(\frac{2m}{h^2}\right)^{3/2}}{N}\times\frac{E_{F_0}^{5/2}}{5/2}$$

$$\bar{E} = \frac{4\pi V\left(\frac{2m}{h^2}\right)^{3/2}\times\frac{2}{5}E_{F_0}^{5/2}}{4\pi V\left(\frac{2m}{h^2}\right)^{3/2}\times\frac{2}{3}E_{F_0}^{3/2}}$$

$$\therefore \qquad \bar{E} = \frac{3}{5}E_{F_0} \qquad \ldots (3.38)$$

3.5 Bloch Theorem (April 16)

- This theorem is applicable when an electron is considered to belong to the crystal as a whole rather than to a particular atom. The one-dimensional Schrödinger's equation for an electron moving in a constant potential V_o is given by

$$\frac{d^2\psi}{dt^2} + \frac{2m}{\hbar^2}(E - V_o)\psi = 0$$

The solutions of above equation are plane waves of the type

$$\psi(x) = e^{\pm ikx}$$

where
$$E - V_o = \frac{\hbar^2 k^2}{2m} = \frac{p^2}{2m} = E_{kin}$$

- For an electron moving in a perfectly periodic potential of the form shown in Fig. 3.8, the Schrödinger's equation is written as

$$\frac{d^2\psi}{dx^2} + \frac{2m}{\hbar^2}[E - V(x)]\psi = 0 \qquad (\because V \text{ depends on } x)$$

- Since the potential is perfectly periodic with period a (where a is a lattice constant), we have

$$V(x) = V(x + a)$$

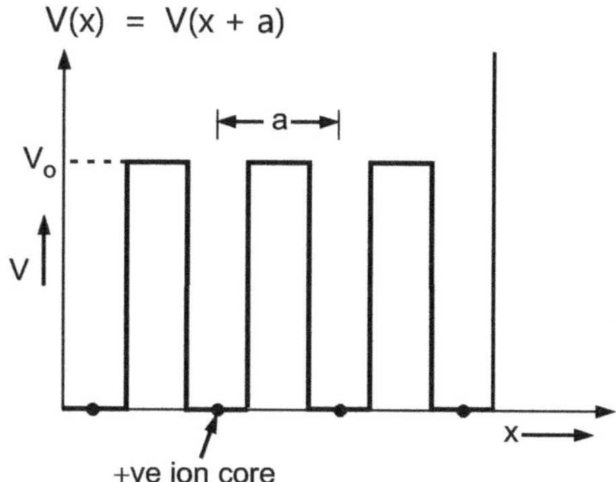

Fig. 3.8 : Ideal periodic potential

- With reference to the solution of Schrödinger's equation, there is an important theorem known as Bloch theorem.

   It states that for the motion of a particle (or electron) moving in a periodic potential, the solutions to the Schrödinger's equations are modulated by a function [$u_k(x)$] having the same periodicity as that of lattice.

- According to this theorem, the solutions are of the form

$$\psi(x) = e^{\pm ikx} u_k(x) \qquad \dots (3.39)$$

   where u_k has periodicity of lattice given by

$$u_k(x) = u_k(x + a) \qquad \dots (3.40)$$

- The wave functions of the type given by equation (3.40) are called the Bloch functions.

- Thus, the eigen functions of the wave equation for a periodic potential are the product of a plane wave exp (ik . **r**) times **a** function **u_k(r)** with the periodicity of the crystal lattice.

- k represents state of motion of electron and k^{th} state corresponds to electron having momentum p = $\hbar$k and a de Broglie wavelength $\lambda = \dfrac{2\pi}{k}$. k can be called as propagation vector. The potential inside the crystal is approximated to the shape of rectangular steps as shown in Fig. 3.8.

3.6 Nearly Free Electron Model

- The behaviour of electron moving in a periodic potential produced by regularly arranged positive ion cores is analysed which is known as nearly free electron model.

- In free electron model, we have seen that the energy of the electron is given by

$$E_k = \frac{\hbar^2 k^2}{2m}$$

For three-dimensional case, $k^2 = k_x^2 + k_y^2 + k_z^2$

$$\therefore \qquad E_k = \frac{\hbar^2}{2m}(k_x^2 + k_y^2 + k_z^2)$$

The allowed energy values are distributed continuously from zero to infinity.

For periodic boundary conditions over a cube of side L,

$$k_x, k_y, k_z = 0; \pm\frac{2\pi}{L}; \pm\frac{4\pi}{L}; \ldots \ldots \qquad \ldots (3.41)$$

The free electron waveforms are of the form

$$\psi_k(\vec{r}) = e^{i(\vec{k} \cdot \vec{r})} \qquad \ldots (3.42)$$

The wave function represents running waves with momentum $P = \hbar k$.

- The band structure of a crystal can be explained by nearly free electron model for which the band electrons are treated as perturbed only by periodic potential of the positive cores. In Bragg's reflection, the reflection of electron waves in crystal is the cause of energy gap.

- To explain Bragg's reflection and origin of energy gap, consider linear solid of lattice constant 'a'.

The Bragg's condition in reciprocal lattice is given by

$$2\vec{k} \cdot \vec{G} + G^2 = 0 \qquad \ldots (3.43)$$

or $\qquad 2kG + G^2 = 0 \qquad\qquad\qquad$ (As $\vec{k} \cdot \hat{} \vec{G} = 0$)

or $\qquad k = \pm\frac{G}{2} \qquad \ldots (3.44)$

But, reciprocal of lattice vector $\vec{G}$ is given by

$$G = \frac{2\pi n}{a} \qquad \ldots (3.45)$$

Substituting equation (3.45) in equation (3.44), we get

$$k = \pm\frac{n\pi}{a} \qquad \ldots (3.46)$$

where n is an integer.

Thus, $k = \pm\dfrac{n\pi}{a}$ is Bragg's reflection condition in (one dimension) this case. The first

reflections and the first energy gap occur at $k = \pm\dfrac{\pi}{a}$ [Refer Fig. 3.9 (a)]. Other energy gaps occur for other values of integer n. The low energy portions of the band structure are shown qualitatively in Fig. 3.9 (a) and (b). Fig. 3.9 (a) is for entirely free electrons and Fig. 3.9 (b) is for electrons that are nearly free with an energy gap at $k = \pm\dfrac{\pi}{a}$.

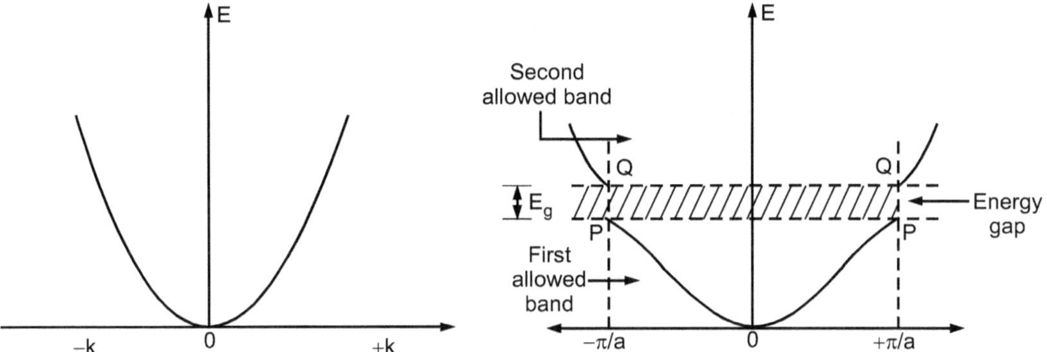

(a) Graph of E versus k for a free electron (b) Plot of E versus k for nearly free electron

Fig. 3.9

- The region in k space between $+\dfrac{\pi}{a}$ and $-\dfrac{\pi}{a}$ is first Brillouin zone of this given lattice. The wave functions at $k = \pm\dfrac{\pi}{a}$ are not travelling waves of the form $e^{\frac{i\pi x}{a}}$ or $e^{-\frac{i\pi x}{a}}$.

- When Bragg's condition is satisfied by a wave vector, a wave travelling in one direction is Bragg's reflected wave and that travelling in opposite direction, each subsequent Bragg's reflection will reverse the direction of travel of wave. So the wave neither travel to the right nor to the left and such a wave is called as **standing wave.**

- Thus, at special values of $k = \pm\dfrac{\pi}{a}, \pm\dfrac{2\pi}{a}, \dots\dots$ the wave functions are made up of equal parts of waves travelling to the right and to the left producing standing waves.

- We can form two different standing waves from the two travelling waves $\psi_1 = e^{+i\pi x/a}$ and $\psi_2 = e^{-i\pi x/a}$, namely

$$\psi(+) = e^{\frac{i\pi x}{a}} + e^{-\frac{i\pi x}{a}}$$

$$\psi(+) = \cos\frac{\pi x}{a} + i\sin\frac{\pi x}{a} + \cos\frac{\pi x}{a} - i\sin\frac{\pi x}{a}$$

or $$\psi(+) = 2\cos\frac{\pi x}{a} \qquad \dots (3.47)$$

Similarly, $$\psi(-) = e^{+\frac{i\pi x}{a}} - e^{-\frac{i\pi x}{a}} = 2i\sin\frac{\pi x}{a} \qquad \dots (3.48)$$

- Both standing waves given by equations (3.47) and (3.48) are composed of equal parts of waves travelling to the right and to the left.

3.7 Fermi Energy and Fermi Level (April 17; Oct. 16, 15)

- The electrons are distributed among the various possible energy levels in accordance with the Pauli's exclusion principle. According to this principle, no two electrons can have all their quantum numbers identical, i.e. each orbital or state can be occupied by at most one electron. This applies to electrons in atoms, molecules or solids.

- In a linear solid, an electron in a conduction electronic state has the quantum numbers n and m_s, where n is the principal quantum number and it takes only positive integer values (i.e. $n = 1, 2, \ldots$) and m_s is magnetic spin quantum number and is equal to $\pm \dfrac{1}{2}$.

  Thus, for each value of n, m_s can have two possible values $+ \dfrac{1}{2}$ or $- \dfrac{1}{2}$. Therefore, each energy level with principal quantum number can accommodate two electrons one with spin up and the other with spin down.

- For example, if there are nine electrons of appropriate spin then in the ground state, first four energy levels would have two electrons each one with spin up and the other with spin down. The fifth level contains the last unpaired electron. Thus, in a system of nine electrons, only 5 levels are occupied and all other energy levels having $n > 5$ will be unoccupied or empty.

- The topmost filled energy level at absolute zero temperature is known as Fermi level and energy corresponding to this level is called Fermi energy for accommodating N electrons at T = 0°K,

$$n_F = \frac{N}{2}$$

- The energy corresponding to n_F is called **Fermi energy**. So energy levels $n > n_F$ will be empty. Therefore, the Fermi level as it divides the filled and empty levels at 0°K. The energy of Fermi level is denoted by E_{F_0}.

- We have discussed the distribution of electrons among the various energy levels at 0°K. The Sommerfeld quantum theory does not allow the condensation of all the electrons into the state of zero energy even at absolute zero temperature.

- Let us study the effect of temperature on occupancy of energy levels. When the temperature is greater than absolute zero, the electrons in the topmost filled energy level may gain energy and get excited to the higher levels. Thus, for T > 0 some of the energy levels below E_{F_0} are empty and above E_{F_0} will be occupied. The probability that a particular state of energy E is occupied at a temperature T is given by Fermi-Dirac distribution. It is given by

$$F(E) \;=\; \frac{1}{1 + \exp\left(\dfrac{E - E_F}{k_B T}\right)} \qquad\qquad \dots (3.49)$$

where F(E) is called the Fermi function,

E is the energy of the level whose occupancy is considered and

E_F is the Fermi energy.

The graph of F(E) versus E is shown in Fig. 3.10.

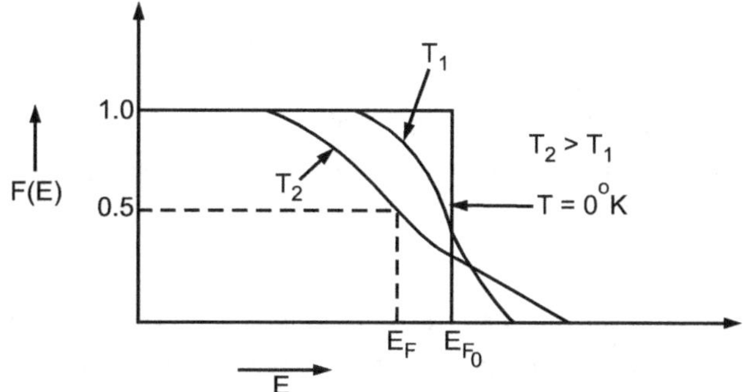

Fig. 3.10 : Fermi distribution function

From Fig. 3.10, $F(E_F) \;=\; \dfrac{1}{2}$.

- For temperature greater than 0°K, the Fermi level can be defined as level where the probability of occupation is $\dfrac{1}{2}$. It must be noted that unlike E_{F_0}, it is not the topmost filled level but it lies between filled level and unoccupied or empty level. The position of Fermi level at a temperature (T) greater than 0°K is not fixed but changes with temperature.

3.8 Hall Effect   (Oct. 17, 16, 15)

- In 1879, G. Hall observed that, when a magnetic field is applied perpendicular to a conductor carrying current, a voltage is developed across the conductor at right angles to both the directions of current flow and that of the magnetic field. This voltage is known as Hall voltage. This phenomenon is known as Hall effect. The value of Hall voltage is found to depend on the magnetic field strength and on the current passed.

- Consider a rectangular metal slab carrying a current density j_x in the positive x-direction and placed in a uniform magnetic field of induction B_z along Z-axis. The Hall field is developed as shown in Fig. 3.11.

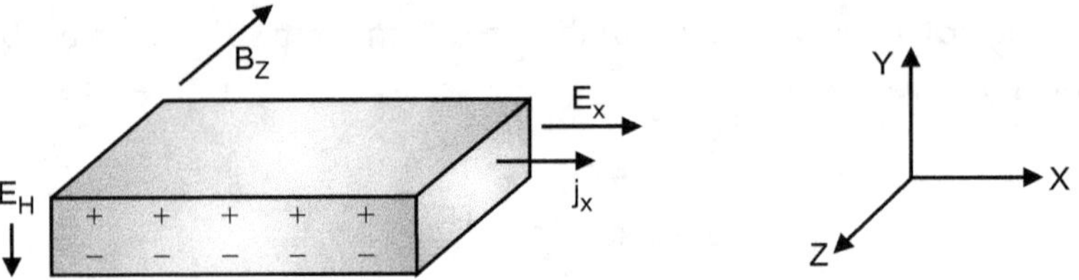

Fig. 3.11 : Hall effect

- In the absence of the magnetic field, the conduction electrons drift with a velocity v_x in the negative x-direction. When the magnetic field is applied, a force called the Lorentz force F_L causes the electrons to deflect downwards as in Fig. 3.11. As a result, electrons accumulate on the lower surfaces, producing a net negative charge there.

- Simultaneously a net positive charge appears on the upper surface, due to the deficiency of electrons. The presence of opposite charges on the opposite faces creates a downward electric field (E_H) called the Hall field.

- The magnitude of Lorentz force F_L producing the charge accumulation in negative y-direction as

$$F_L = e v_x B_z \qquad \text{... (3.50)}$$

- The field created by the surface charges produces a force which opposes this Lorentz force. The accumulation continues until the Hall force completely cancels the Lorentz force. Thus in equilibrium,

$$F_L = F_H \qquad \text{... (3.51)}$$
$$e E_H = e v_x B_z$$
$$E_H = v_x B_z$$
$$\therefore \quad v_x = \frac{E_H}{B_z} \qquad \text{... (3.52)}$$

The equation for current density j_x is

$$j_x = -n e v_x \qquad \text{... (3.53)}$$

Using equation (3.52), we have

$$j_x = -n e \left(\frac{E_H}{B_z} \right) \qquad \text{... (3.54)}$$

$$E_H = -\left(\frac{1}{n e} \right) B_z j_x \qquad \text{... (3.55)}$$

- This shows that the Hall field is proportional to both the magnetic field and the current density. The constant of proportionality in equation (3.54) is known as Hall coefficient and is defined by

$$R_H = -\frac{1}{n e} = \frac{E_H}{j_x B_z} \qquad \text{... (3.56)}$$

- It is inversely proportional to the density of charge carrier n.

- Hall constant or coefficient is defined as ratio of electric field strength produced per unit current density to the transverse magnetic field.

$$\text{Unit of } R_H = \frac{\text{Volt/m}}{\text{Amp-Weber/m}^2} = Vm^3 \, A^{-1} \, Wb^{-1}$$

- In the monovalent metals R_H is negative, the predominant charge is electron and vice-versa. For example for Al, V is $-0.31 \times 10^{-10} \, Vm^3 \, A^{-1} \, Wb^{-1}$. In some of the metals, there is band overlap and R_H can be positive (e.g. in Zinc and Cadmium).

Mobility and Hall angle :

The mobility (μ) is defined as the *velocity acquired by the current carrying particles per unit electric field.*

$$\mu = \frac{v_x}{E_x}$$

$$\therefore \qquad v_x = \mu E_x \qquad \qquad \text{... (3.57)}$$

where, E_x is the applied electric field due to which current flows through a conductor.

Substituting in equation (3.52), we get

$$\mu E_x = \frac{E_H}{B_z}$$

$$\therefore \qquad \mu = \frac{E_H}{E_x \, B_z} \qquad \qquad \text{... (3.58)}$$

Substituting from equation (3.56), we get

$$\mu = \frac{j_x \, B_z}{E_x \, B_z} \, R_H$$

$$\mu = \left(\frac{j_x}{E_x}\right) R_H$$

$$\mu = \sigma R_H \qquad \qquad \text{... (3.59)}$$

where $\sigma = \dfrac{j_x}{E_x}$ is the electrical conductivity of specimen.

From equation (3.58), we get

$$\frac{E_H}{E_x} = \mu B_z = \phi \qquad \qquad \text{... (3.60)}$$

where ϕ is called Hall angle.

The measurement of Hall voltage helps us to know the following :

1. The sign of current carrying charges an be determined

2. The charge density can be calculated.

3. The mobility of charge carriers can be obtained from measurement of Hall voltage.

3.9 Origin of Energy Gap

- In above section, we have seen that, first reflections and first energy gap occur at $k = \pm \pi/a$.

- In between $k = +\dfrac{\pi}{a}$ and $k = -\dfrac{\pi}{a}$, the wave functions are the travelling waves having equations of the form $e^{i\pi x/a}$ and $e^{-i\pi x/a}$ of the free electrons. The two standing waves given by equations (3.57) and (3.58) pile up electrons at different regions and so the two waves have different values of the potential energy. This is the origin of energy gap.

  The probability density of a particle (or charge density) is given by

  $$\rho = \psi\psi^* = |\psi|^2$$

- For pure travelling waves having equation $\psi = e^{ikx}$, we have $|\psi|^2 = e^{ikx}\, e^{-ikx} = 1$, so the charge density is constant. However, the charge density is not constant for linear combination of plane waves.

- For standing wave $\psi(+)$, the charge density is

  $$\rho(+) = |\psi(+)|^2 \propto \cos^2 \frac{\pi x}{a} \qquad\qquad \dots (3.61)$$

- At $x = 0, a, 2a \ldots$ $\rho(+)$ is maximum. Thus, standing wave $\psi(+)$ piles up negative charge (electron) on the positive ion cores centred at $x = 0, a, 2a, \ldots$ as shown in Fig. 3.12 (a) where the potential energy is lowest.

- On the other hand, for standing wave $\psi(-)$, the charge density is

  $$\rho(-) = |\psi(-)|^2 \propto \sin^2 \frac{\pi x}{a} \qquad\qquad \dots (3.62)$$

- From above equation, $x = 0, a, 2a, \ldots$, we get $\rho(-) = 0$. Thus, the standing wave $\psi(-x)$ concentrates electrons or negative charge away from the positive ion cores situated at $x = 0, a, 2a, \ldots$. Fig. 3.12 (b) shows the electron or charge concentration for the standing waves represented by $\psi(+)$ and $\psi(-)$ and for travelling waves.

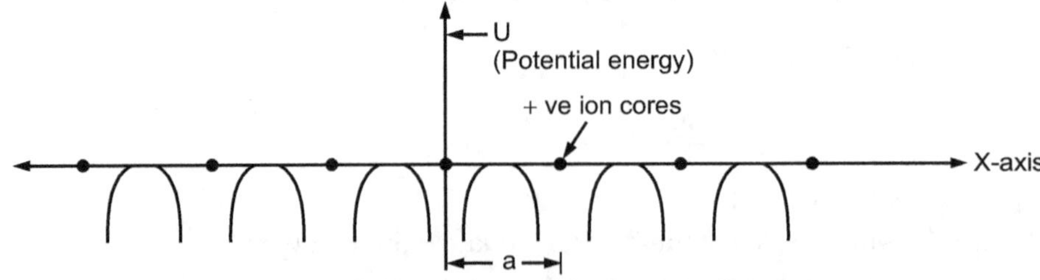

(a) Variation of potential energy of free electrons

in the field of positive ion core of linear lattice

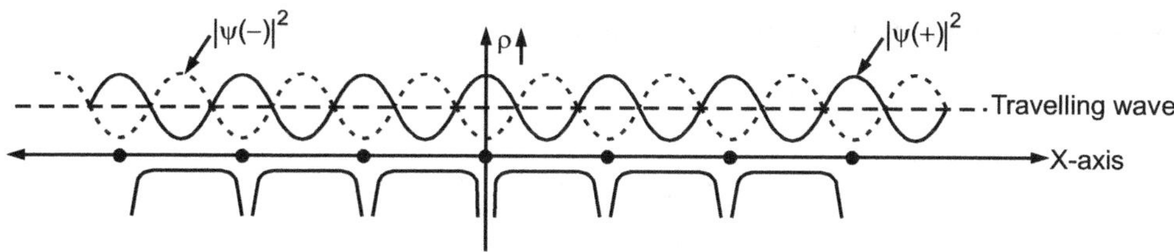

(b) Distribution of charge density in the lattice for $|\psi(+)|^2$, $|\psi(-1)|^2$ for travelling waves

Fig. 3.12

- From equations (3.61) and (3.62), if we calculate average potential energy over these regions, we find that the potential energy of $\rho(+)$ or $|\psi(+)^2|$ is lower than that of travelling wave whereas the potential energy of $\rho(-)$ is higher than travelling wave. If the energies of $\rho(+)$ and $\rho(-)$ differ by some value say E_g, we have an energy gap of width E_g. In Fig. 3.12 (b), just below the energy gap at point P, the wave function is $\psi(+)$ and just above the energy gap at point Q the wave function is $\psi(-)$.

Let us calculate the magnitude of the energy gap.

The wave functions at the zone boundary $k = \dfrac{\pi}{a}$ and $k = -\dfrac{\pi}{a}$ are $\sqrt{2}\cos\dfrac{\pi x}{a}$ and $\sqrt{2}\sin\dfrac{\pi x}{a}$ respectively, where $\sqrt{2}$ is a normalization constant.

The potential energy of an electron in the crystal at point x is given by

$$U(x) = U\cos\frac{2\pi x}{a} \qquad \text{... (3.63)}$$

The first order energy difference E_g between the two standing wave states is

$$E_g = \int_0^1 U(x)\,[|\psi(+)|^2 - |\psi(-)|^2]\,dx$$

Using wave functions $\psi(+) = \sqrt{2}\cos\dfrac{\pi x}{a}$ and $\psi(-) = \sqrt{2}\sin\dfrac{\pi x}{a}$ in above equation, we get

$$E_g = 2\int U\cos\left(\frac{2\pi x}{a}\right)\left(\cos^2\frac{\pi x}{a} - \sin^2\frac{\pi x}{a}\right)dx$$

or $\qquad\qquad\qquad E_g = U \qquad\qquad\qquad\qquad \text{... (3.64)}$

Thus, the energy gap is equal to the amplitude of the Fourier component of the crystal potential.

3.10 Effective Mass of an Electron

- According to wave mechanical theory of particles, a particle moving with velocity V is equivalent to a wave packet with a group velocity V_g.
- The group velocity of wave packet is given by

$$V = V_g = \frac{d\omega}{dK} \qquad\qquad \text{... (3.65)}$$

where ω is angular frequency of the de-Broglie waves.

- The energy E of a moving particle is given by

$$E = h\nu$$

or

$$E = \hbar\omega \qquad (\because \omega = 2\pi\nu)$$

Differentiating above equation w.r.t. K, we get

$$\frac{dE}{dK} = \hbar \frac{d\omega}{dK}$$

or

$$\frac{d\omega}{dK} = \frac{1}{\hbar} \frac{dE}{dK} \qquad \ldots (3.66)$$

Substituting equation (3.64) in equation (3.63), we get

$$V_g = \frac{1}{\hbar} \frac{dE}{dK}$$

From above equation it is clear that group velocity V_g depends on $\dfrac{dE}{dK}$.

The variation of energy E with K in the first Brillouin zone is shown in Fig. 3.13.

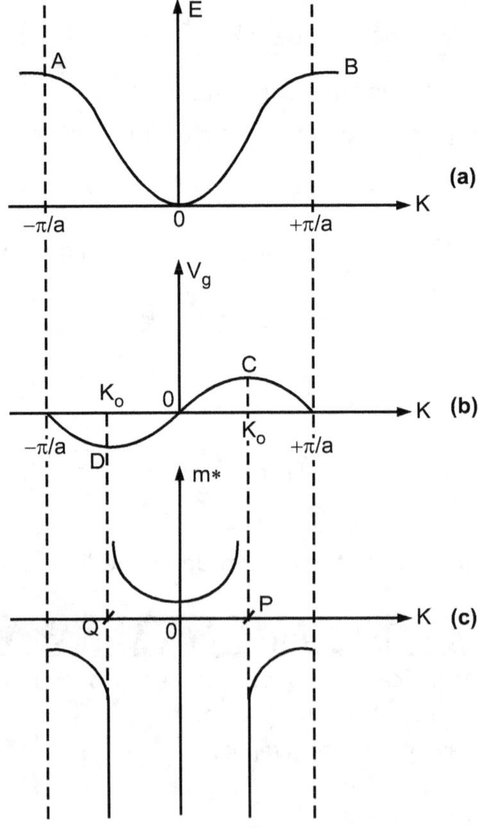

Fig. 3.13 : Variation of E, V_g and m* with K in one dimension

In Fig. 3.13 (a), $\dfrac{dE}{dK}$ at the zone boundary i.e. at points A and B is zero. Thus, at Brillouin zone boundary i.e. at $K = \pm \dfrac{\pi}{a}$, the group velocity is $V_g = 0$, because $\dfrac{dE}{dK} = 0$. So at the Brillouin zone boundary, we cannot have propagating waves but only stationary waves are formed due to back and forth Bragg's reflection. In between $-\dfrac{\pi}{a}$ to $+\dfrac{\pi}{a}$ the slope $\dfrac{dE}{dK}$ of the curve [Refer Fig. 3.13 (a)] is not constant but changes with K. From the slope of this curve, curve V_g versus K is obtained [Refer Fig. 3.13 (b)]. This curve indicates that the velocity of electron is zero for $K = 0$ and for $K = \pm \pi/a$. For $K = K_0$, the group velocity is maximum. The point on the K-axis, where V_g is maximum is called the point of inflexion. Now, let us study the effect of applied electric field ε on the motion of an electron present in the Brillouin zone or energy band.

If the electron field ε acts on electron for a time dt producing displacement dx, then work done by electric field is

$$d\omega = F\,dx = e\varepsilon\,dx \qquad\qquad (\because F = e\varepsilon)$$

$\therefore$ The increase in energy of the electron is

$$dE = e\varepsilon\,dx$$

But,
$$V_g = \frac{dx}{dt}$$

$\therefore$
$$dE = e\varepsilon\,V_g\,dt$$

Using group velocity of wave packet $V_g = \dfrac{1}{\hbar}\dfrac{dE}{dK}$ in above equation, we get

$$dE = \frac{e\varepsilon}{\hbar}\frac{dE}{dK}\,dt$$

or
$$\frac{dE}{dK}\,dK = \frac{e\varepsilon}{\hbar}\frac{dE}{dK}\,dt$$

$\therefore$
$$\frac{dK}{dt} = \frac{e\varepsilon}{\hbar}$$

or
$$e\varepsilon = \hbar\frac{dK}{dt} \qquad\qquad\qquad \ldots (3.67)$$

The acceleration of the wave packet is given by

$$\frac{d}{dt}V_g = \frac{dV_g}{dK}\cdot\frac{dK}{dt}$$

But,
$$V_g = \frac{1}{\hbar}\frac{dE}{dK}$$

$$\therefore \quad \frac{dV_g}{dt} = \frac{d}{dK}\left(\frac{1}{\hbar}\frac{dE}{dK}\right)\frac{dK}{dt}$$

$$\text{or} \quad \frac{dV_g}{dt} = \frac{1}{\hbar}\frac{d^2E}{dK^2}\frac{dK}{dt}$$

Therefore, acceleration of wave packet $= \dfrac{1}{\hbar}\dfrac{d^2E}{dK^2}\dfrac{dK}{dt}$ ... (3.68)

Substituting equation (3.67) in equation (3.68), we get

$$\text{Acceleration} = \frac{1}{\hbar}\frac{d^2E}{dK^2}\cdot\frac{e\varepsilon}{\hbar}$$

$$\text{or} \quad a = \frac{e\varepsilon}{\hbar^2}\frac{d^2E}{dK^2} \qquad \qquad ... (3.69)$$

But for free electron of mass m,

$$a = \frac{e\varepsilon}{m} \qquad \qquad (\text{Since } F = e\varepsilon \text{ or } ma = e\varepsilon) \; ... (3.70)$$

Equating equations (3.69) and (3.70), we get

$$\frac{e\varepsilon}{m} = \frac{e\varepsilon}{\hbar^2}\frac{d^2E}{dK^2}$$

From above equation, it is clear that an electron moving in a crystal has an effective mass m* given by

$$m^* = \frac{\hbar^2}{\left(\dfrac{d^2E}{dK^2}\right)} \qquad \qquad ... (3.71)$$

This equation indicates that effective mass of an electron is not constant but depends on the value of $\dfrac{d^2E}{dK^2}$. The variation of effective mass m* with K is shown in Fig. 3.13 (c). The effective mass m* is positive in the inner half of the Brillouin zone and negative in the outer half portion. The mass m* becomes infinite at the inflexion points of the E(K) curve.

To give physical interpretation of factor m*, let us introduce factor f_k given by

$$f_k = \frac{m}{m^*} = \frac{m}{\hbar^2}\left(\frac{d^2E}{dK^2}\right) \qquad \qquad ... (3.72)$$

If $f_k = 1$, the electron behaves totally as free electron and if $f_k < 1$ i.e. $m^* > m$, the electron behaves as a heavy particle.

When electron is in the lower part of the band (i.e. in the region between points P and Q in Fig. 3.13 (c), m* is positive, where an increase in energy is accompanied by an increase in the momentum. On the other hand, when electron is in the upper band (or outer half zone), m* assumes negative value and increase in the energy amounts to decrease in the value of momentum.

Limitations of Sommerfeld Theory

- The Sommerfeld's free electron theory described earlier, was successful in explaining the various electronic and thermal properties of metals such as specific heat, paramagnetism, etc. However, this theory is incapable of explaining why some solids are good conductors, some are semiconductors and others are insulators.

- The most important feature of Sommerfeld theory is that it destroys the concept or assumption that all free electrons are conduction electrons. But, only few electrons whose energies lie near the Fermi levels take part in conduction and these free electrons are called conduction electrons. The failure of Sommerfeld's theory is due to the oversimplified assumption that conduction electrons move freely in a constant potential within the volume of solid. Infact, the potential due to positive ion cores is not constant but changes with position of electron in a crystal.

- The periodic arrangement of positive ions in crystal produces a potential which changes periodically. The approximation of a constant potential inside the metal is correct to certain extent. So this is the reason why free electron theory explain electrical properties of metals. But this will not be valid for all solids.

- The periodic potential described above forms the basis of the band theory.

3.11 Energy Bands in a Solid

- A solid is formed when atoms bond together. Atom consists of positively charged nucleus at its centre and electrons are revolving round the nucleus in certain fixed orbits called stationary or permitted orbits. The electrons in the outermost orbit are called as valence electrons. Each electron in the solid occupies a different position inside the solid and hence, no two electrons inside the solid have exactly the same pattern of surrounding charges. All electrons in the solid belonging to the first orbit have slightly different energy levels because no two electrons have exactly the same pattern of surrounding charges. Since there are billions of first orbit electrons, the group of slightly different energy levels corresponding to all these first orbit electrons form a band. In a similar manner, electrons inside the solid belonging to the second orbit form the second energy band and so on.

- The group of slightly different energy levels of all electrons belonging to the valence orbit (outermost orbit) form a band called as a valence band. This is the band, in which valence electrons from each atom of the solid will be located. All the permissible energy levels in the valence band are occupied by electrons. These electrons are possible carriers of electricity. The upper energy band (Refer Fig. 3.14) called as conduction band is empty of electrons because it corresponds to the unoccupied higher levels of electrons of atoms. The energy region or gap between conduction band and valence band is called the forbidden gap since, no electrons with such energies exist in the solid.

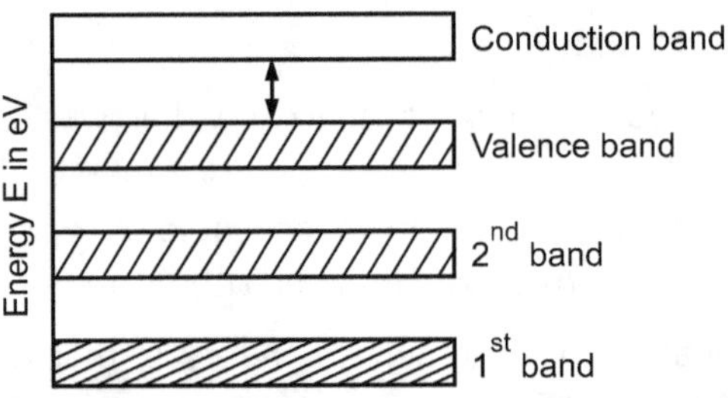

Fig. 3.14 : Energy levels for solid (at 0°K temperature)

3.12 Distinction Between Metals, Insulators and Semiconductors

(Oct. 15)

- To distinguish between metals, semiconductors and insulators, consider a one-dimensional solid having lattice constant 'a'. Consider a particular energy band filled with electrons upto certain value K_1 ($K_1 < \pi/a$) at T = 0 as shown in Fig. 3.15.

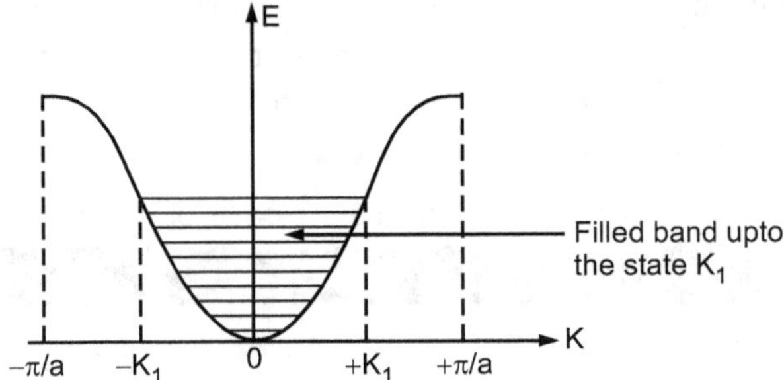

Fig. 3.15 : Energy band (first Brillouin zone)

- Let us introduce the term

$$f_K = \frac{m}{m^*} = \frac{m}{\hbar^2} \frac{d^2E}{dK^2}$$

which represent the extent upto which an electron in the K-state behaves as a free electron and hence takes part in conduction.

- To determine the total effective number of free electrons in the given energy band, we have to sum up overall possible $f_{K's}$ corresponding to various occupied states in the band.

$$\therefore \qquad N_{eff} = \Sigma f_K \qquad\qquad ... (3.73)$$

- The number of states in the range dK for a one-dimensional lattice of length L is given by

$$dn = \frac{L}{2\pi} dK \qquad\qquad ...(3.74)$$

- As each state is occupied by maximum number of two electrons, the effective number of electron present in the band between $-K_1$ to K_1 (shaded region) becomes

$$N_{eff} = 2\frac{L}{2\pi} \int_{-K_1}^{K_1} f_K \, dK$$

Using $f_K = \dfrac{m}{\hbar^2}\dfrac{d^2E}{dK^2}$ in above equation, we get

$$N_{eff} = \frac{L}{\pi} \int_{-K_1}^{+K_1} \frac{m}{\hbar^2}\frac{d^2E}{dK^2}\, dK$$

or

$$= \frac{2mL}{\pi\hbar^2} \int_0^{K_1} \frac{d}{dK}\left(\frac{dE}{dK}\right) dK$$

$$N_{eff} = \frac{2mL}{\pi\hbar^2}\left[\frac{dE}{dK}\right]_{K=K_1} \qquad \ldots (3.75)$$

From equation (3.75), two important conclusions can be drawn.

(i)   The effective number of free electrons in completely filled band is zero. This is because, if the band is completely filled (from $-\pi/a$ to $+\pi/a$ in Fig. 3.16) then $\dfrac{dE}{dK}$ is zero at the top of band.

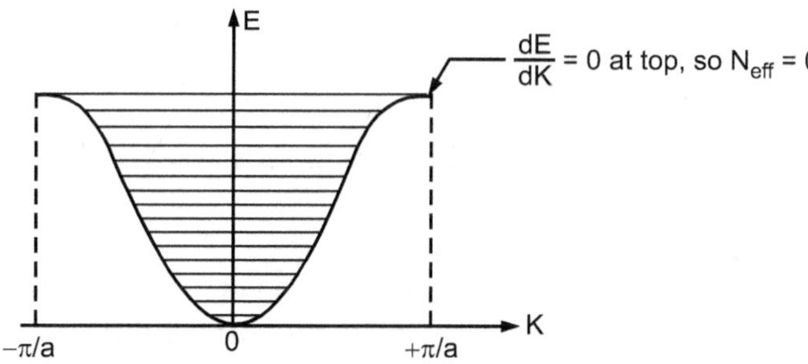

Fig. 3.16 : Completely filled band

(ii)   The effective number of electrons attains a maximum value when the band is filled upto the point of inflexion as shown in Fig. 3.17. This is because at $K = \pm K_a$, $\dfrac{dE}{dK}$ is maximum.

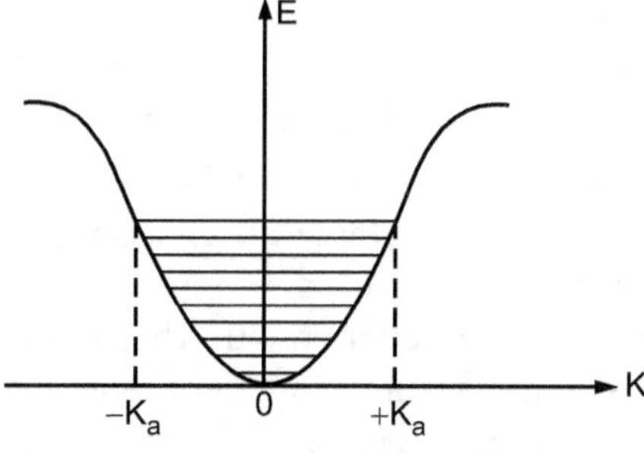

Fig. 3.17 : The band filled upto the point of inflexion

- From above discussion, we can say that a solid having a certain number of energy bands completely filled and all others are absolutely vacant are an insulator. The band formed by a series of energy levels containing valence electrons is called valence band and the one on the top of valence band is called conduction band. The energy gap between the valence band and conduction band is called band gap.

- A solid having partially filled energy band has a metallic character. It should be noted that the situation shown in Fig. 3.18 is strictly at 0°K temperature. A solid having partially or partly filled energy bands has effective number of free electrons which can take part in conduction.

- At temperature greater than 0°K, some electrons from the valence band jump to the next higher empty band (conduction band) where they take part in the conduction process.

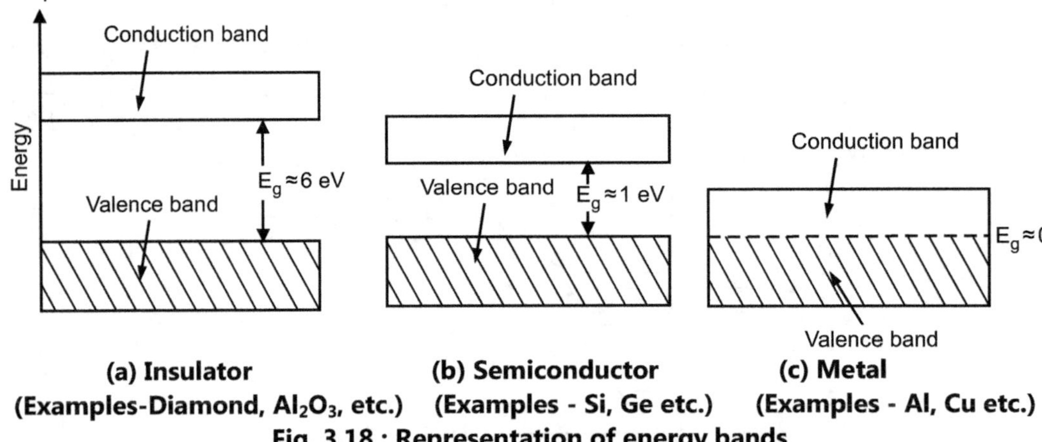

(a) Insulator   **(b) Semiconductor**   **(c) Metal**
(Examples-Diamond, Al$_2$O$_3$, etc.)   (Examples - Si, Ge etc.)   (Examples - Al, Cu etc.)
Fig. 3.18 : Representation of energy bands

- For conduction to occur, the electrons in the valence band must get sufficient energy to overcome the band gap. Since in insulator like diamond, the band gap is about 6 eV. It is impossible to excite the electrons from valence band to conduction band. The band gap in certain materials like Si, Ge is small ($\approx$ 1 eV) at 0°K. So these materials are also insulators at 0°K. But even at room temperature, few electrons in the valence band acquire energy and get excited to conduction band. So at room temperature, these materials show small conductivity and hence they are called semiconductors. Fig. 3.18 shows energy bands for (a) insulators, (b) semiconductors and (c) metals.

- In case of metals, there will be overlapping of valence band and conduction band. A large number of conduction electrons are available in them even at low temperature.

- It is important to note that the conductivity of semiconductor increases with increase in temperature. This is because with increase in temperature, more and more electrons jump from valence band to conduction band. The conductivity of metals, however, decreases with increase in temperature. This is because with increase in temperature, more and more phonons are excited which scatter conduction electrons and so reduces mobility of electrons.

Solved Examples

Example 3.1 : *The Fermi energy of copper is 7.1 eV. Assuming that it is the maximum kinetic energy of electrons in copper, find the number of atoms per unit volume in copper.*

Solution : Fermi energy E_F is   (Oct. 16)

$$E_F = \frac{h^2}{2m}\left(\frac{3N}{8\pi V}\right)^{2/3}$$

$$\therefore \qquad N = \frac{8}{3}\pi V\left(\frac{2mE_F}{h^2}\right)^{3/2}$$

The number of electrons per unit volume is

$$\frac{N}{V} = \frac{8}{3}\pi\left(\frac{2mE_F}{h^2}\right)^{3/2}$$

A copper atom contributes one electron to the Fermi electron gas.

$m = 9.1 \times 10^{-31}$ kg, $E_F = 7.1$ eV $= 7.1 \times 1.6 \times 10^{-19}$ joule, $h = 6.62 \times 10^{-34}$ joule-sec.

$$\frac{N}{V} = \frac{8}{3}\pi\left[\frac{2 \times 9.1 \times 10^{-31} \times 7.1 \times 1.6 \times 10^{-19}}{(6.62 \times 10^{-34})^2}\right]^{3/2}$$

$$\frac{N}{V} = \frac{8}{3}\pi\,(4.7177 \times 10^{18})^{3/2}$$

$$\frac{N}{V} = \mathbf{8.58 \times 10^{28}\ atoms/m^3} \qquad\qquad \textbf{... Ans.}$$

Example 3.2 : *Calculate Fermi energy of potassium from the following data :*

Density of potassium = 0.86 g/cm³

Atomic weight = 39.202

Mass of the electron = 9.1 ×10⁻²⁸ g

Planck's constant, h = 6.62 ×10⁻²⁷ ergs-sec.

1 eV = 1.602 ×10⁻¹² ergs.

Solution : The number of electrons per cm³

$$\frac{N}{V} = \frac{\text{Density} \times \text{Avogadro's number}}{\text{Atomic weight}}$$

$$\frac{N}{V} = \frac{0.86 \times 6.023 \times 10^{23}}{39.202}$$

$$\frac{N}{V} = 1.321 \times 10^{22}$$

Fermi energy E_F is

$$E_F = \frac{h^2}{2m}\left(\frac{3}{8\pi}\frac{N}{V}\right)^{2/3}$$

$$= \frac{(6.62 \times 10^{-27})^2}{2 \times 9.1 \times 10^{-28}}\left(\frac{3}{8 \times 3.14} \times 1.321 \times 10^{22}\right)^{2/3}$$

$$= (2.408 \times 10^{-26}) \times (1.355 \times 10^{14})$$

$$= 3.263 \times 10^{-12} \text{ ergs}$$

$$= \frac{3.263 \times 10^{-12}}{1.602 \times 10^{-12}} \text{ eV}$$

$$\therefore \qquad E_F = \textbf{2.037 eV} \qquad\qquad\qquad \textbf{... Ans.}$$

Example 3.3 : *Calculate the number of energy states available for the electrons in a cubical box of side 0.05 cm lying below an energy of 1 eV.*

Given : m = 9.1 × 10⁻³¹ kg, h = 6.62 × 10⁻³⁴ joule-sec.

Solution :
$$Z(E)\,dE = 4\pi V \left(\frac{2m}{h^2}\right)^{3/2} E^{1/2}\,dE$$

The number of energy states below 1 eV is

$$\int_0^E Z(E)\,dE = 4\pi V \left(\frac{2m}{h^2}\right)^{3/2} \int_0^E E^{1/2}\,dE$$

$$= 4\pi V \left(\frac{2m}{h^2}\right)^{3/2} \left[\frac{2}{3} E^{3/2}\right]_0^{1eV}$$

where, m = 9.1×10^{-31} kg, V = 1.25×10^{-10} m³, E = $1 \times 1.6 \times 10^{-19}$ joule

The number of energy states is

$$= 4 \times 3.14 \times 1.25 \times 10^{-10} \times \left[\frac{2 \times 9.1 \times 10^{-31}}{(6.62 \times 10^{-34})^2}\right]^{3/2} \times \left[\frac{2}{3}(1.6 \times 10^{-19})\right]^{3/2}$$

$$= 4 \times 3.14 \times 1.25 \times 10^{-10} \times 8.463 \times 10^{54} \times \frac{2}{3} \times 6.4 \times 10^{-29}$$

$$= \textbf{5.669} \times \textbf{10}^{\textbf{17}} \qquad\qquad\qquad \textbf{... Ans.}$$

Example 3.4 : *Evaluate the temperature at which there is one percent probability that a state, with an energy 0.4 eV above the Fermi energy, will be occupied by an electron.*

(Given : $K_B = 1.38 \times 10^{-23}$ joule/kelvin)   **(April 17, Oct. 15)**

Solution :
$$F(E) = \frac{1}{1 + \exp\left[\dfrac{(E - E_F)}{K_B T}\right]}$$

$$E = E_F + 0.4$$

$$F(E) = \frac{1}{100} = \frac{1}{1 + \exp(0.4/K_B T)}$$

$$100 = 1 + \exp(0.4/K_B T)$$

$$\exp(0.4/K_B T) = 99$$

$$\frac{0.4}{K_B T} = 2.303 \times \log_{10} 99$$

$$K_B T = \frac{0.4}{2.303 \times \log_{10} 99}$$

$$K_B T = 0.087 \text{ eV}$$

$$T = \frac{0.087}{K_B}$$

$$T = \frac{0.087}{1.38 \times 10^{-23}}$$

$$\therefore \qquad T = \textbf{1008.69 Kelvin} \qquad\qquad \textbf{... Ans.}$$

Example 3.5 : *Calculate the Hall coefficient of sodium based on free electron model. Sodium has BCC structure and the side of the cube is 4.2 A°.*

(Given : Charge on an electron = 1.6 $\times 10^{-19}$ C)

Solution : For sodium, side of cube, a = 4.2 A° = 4.2×10^{-10} m

BCC structure of crystal indicates 2 atoms per unit cell. Hence, the number of electrons per unit volume,

$$n = \frac{2}{a^3}$$

$$= \frac{2}{(4.2 \times 10^{-10})^3}$$

$$= 2.699 \times 10^{28} \text{ /m}^3$$

The Hall coefficient is

$$R_H = -\frac{1}{ne}$$

$$= -\frac{1}{2.699 \times 10^{28} \times 1.6 \times 10^{-19}}$$

$$= \textbf{− 2.3156} \times \textbf{10}^{-10} \textbf{ m}^3\textbf{/coulomb} \qquad\qquad \textbf{... Ans.}$$

Example 3.6 : *An n-type semiconductor (Ge) has a donor density of $10^{15}/cm^3$. It is arranged in a Hall effect experiment where magnetic field $B_z = 0.5$ Wb/m^2 is applied and a current density of $j_x = 500$ A/m^2 results. What will be the Hall voltage if the specimen is 4 mm thick ?*

Solution :
$$j_x = 500 \text{ A/m}^2 = 500 \times 10^{-4} \text{ A/cm}^2$$
$$B_z = 0.5 \text{ Wb/m}^2 = 0.5 \times 10^{-4} \text{ Wb/cm}^2$$
$$n = 10^{15} \text{ /cm}^3$$
$$e = 1.6 \times 10^{-19} \text{ coulomb}$$
$$d = 4 \text{ mm} = 0.4 \text{ cm}$$
$$R_H = -\frac{1}{ne} = \frac{E_H}{j_x B_z}$$
$$E_H = -\frac{j_x B_z}{ne}$$
$$V_H = E_H \cdot d \quad \text{(d is the thickness)}$$

From previous equations, we get

$$V_H = -\left(\frac{j_x B_z}{ne}\right) d$$
$$= -\left(\frac{500 \times 10^{-4} \times 0.5 \times 10^{-4}}{10^{15} \times 1.6 \times 10^{-19}}\right) \times 0.4$$
$$= -(0.015625) \times 0.4$$
$$= -0.00625$$
$$= -6.25 \times 10^{-3} \text{ volts} \qquad \textbf{... Ans.}$$

Example 3.7 : *A sample of Si is doped with 10^{17} phosphorous atoms per cm^3. What is its resistivity ? What is the expected Hall voltage in a sample of 200 μm thickness if the current density is 1 A/cm^2 and magnetic field of 1×10^{-5} Wb/cm^2 is applied perpendicular to the direction of current flow.*

(Given : Mobility = 600 cm^2/volt-sec.)

Solution : n = 10^{17} electrons/cm^3, μ = 600 cm^2/volt-sec., $B_z = 1 \times 10^{-5}$ Wb/cm^2,

d = 200 μm, $j_x = 1$ A/cm^2

Conductivity,
$$\sigma = \frac{\mu}{R_H} = \mu n e$$
$$= 600 \times 10^{17} \times 1.6 \times 10^{-19}$$
$$= 9.6 \ \Omega\text{-cm}$$

Resistivity, $\qquad \rho = \dfrac{1}{\sigma}$

$$= \dfrac{1}{9.6}$$

$$= 0.104 \ \Omega\text{-cm}$$

Hall coefficient, $\qquad R_H = -\dfrac{1}{ne}$

$$= -\dfrac{1}{10^{17} \times 1.6 \times 10^{-19}}$$

$$= -62.5 \ \text{cm}^3/\text{coulomb}$$

Hall voltage, $\qquad V_H = E_H \, d$

$$= (j_x \, B_z \, R_H) \, d$$

$$= 1 \times 1 \times 10^{-5} \times (-62.5) \times (2 \times 10^{-2})$$

$$= 12.5 \times 10^{-6} \ \text{V}$$

$$= \textbf{12.5 } \boldsymbol{\mu}\textbf{V} \qquad\qquad \textbf{... Ans.}$$

Example 3.8 : *In a Hall effect experiment on zinc, a potential of 4.5 μV is developed across a foil of thickness 0.02 mm when a current of 1.5 A is passed in a direction perpendicular to a magnetic field of 2.0 T. Calculate :*

(a)   The Hall coefficient for zinc.

(b)   The electron density. $\qquad\qquad\qquad\qquad\qquad\qquad\qquad$ **(April 16)**

Solution : (a) $\qquad V_H = 4.5 \times 10^{-6}$ V, $t = 2 \times 10^{-5}$ m, $i = 1.5$ A, $B = 2$T

$$R_H = \dfrac{V_H t}{iB}$$

$$= \dfrac{(4.5 \times 10^{-6}) \, (2 \times 10^{-5})}{(1.5) \, (2)}$$

$$= 0.3 \times 10^{-10} \ \text{m}^3 \ \text{C}^{-1} \qquad\qquad \textbf{... Ans.}$$

(b) $\qquad\qquad\qquad n = \dfrac{1}{R_H e}$

$$= \dfrac{1}{(0.3 \times 10^{-10}) \, (1.6 \times 10^{-19})}$$

$$= \textbf{2.08} \times \textbf{10}^{\textbf{29}} \ \textbf{m}^{-3} \qquad\qquad \textbf{... Ans.}$$

Example 3.9 : *Calculate for silver the energy at which the probability that a conduction electron state will be occupied is 90%. Assume E_F = 5.52 eV for silver and temperature T = 800 K.*

Solution :

$$F(E) = \frac{1}{1 + \exp\left(\dfrac{E - E_F}{k_B T}\right)}$$

Substituting $\quad kT = 5.52 \times 10^{-5} \times 800 = 0.04416$ eV

Solving for ΔE, we get $\Delta E = E - E_F = -2.2 \times 0.04416 = -0.097$

$\therefore \qquad\qquad E = 5.52 - 0.10 = 5.42$ eV

$$F(E) = \mathbf{0.9} \qquad\qquad\qquad\qquad \textbf{... Ans.}$$

Summary

1. P. Drude and Lorentz proposed the free electron theory of metals in 1990.

2. The path between the collisions is free path and the average distance travelled by the electron between two collisions is called mean free path

3. Conductivity $\sigma = \dfrac{ne^2 L}{mv_{rms}}$.

4. Wiedemann-Franz law : $\dfrac{K}{\sigma} = \dfrac{3}{2} \dfrac{k_B^{\,2}}{e^2} T$.

5. The ratio $K/\sigma T$ is a constant and known as the Lorentz number.

6. The quantum free electron theory of Sommerfeld has three basic assumptions :

    (i) Valence electrons of constituent atoms move freely throughout the volume of the metal like gas molecules in a vessel.

    (ii) The free electrons obey Pauli's exclusion principle.

    (iii) The free electrons in a solid move as if it is a system of waves.

7. Energy of electron in one dimension $E_n = \dfrac{n^2 \pi^2 \hbar^2}{2mL^2}$.

8. The Fermi energy E_{F_0} is defined as the energy of the topmost filled level in the ground state of the N electron system.

9. Density of states in one dimension, $Z(E) = 2 \times \dfrac{L}{4\pi} \left(\dfrac{2m}{\hbar^2}\right)^{1/2} E^{-1/2}$.

10. In three dimension $E_n = \dfrac{\hbar^2 \pi^2}{2mL^2} n^2$ and density of states, $Z(E)\, dE = 2\pi V \left(\dfrac{2m}{h^2}\right)^{3/2} E^{1/2}\, dE$.

11. Bloch theorem states that for the motion of a particle (or electron) moving in a periodic potential, the solutions to the Schrödinger's equations are modulated by a function $[u_k(x)]$ having the same periodicity as that of lattice.

12. Fermi-Dirac distribution given by $F(E) = \dfrac{1}{1 + \exp\left(\dfrac{E - E_F}{k_B T}\right)}$ where $F(E)$ is called the Fermi function.

13. Hall observed that, when a magnetic field is applied perpendicular to a conductor carrying current, a voltage is developed across the conductor at right angles to both the directions of current flow and that of the magnetic field. This voltage is known as Hall voltage. This phenomenon is known as Hall effect.

14. The mobility (μ) is defined as the velocity acquired by the current carrying particles per unit electric field. $\mu = \dfrac{v_x}{E_x}$ **(April 17)**

15. The energy gap is equal to the amplitude of the Fourier component of the crystal potential.

16. Effective mass of electron $m^* = \dfrac{\hbar^2}{\left(\dfrac{d^2E}{dK^2}\right)}$.

17. In case of metals, there will be overlapping of valence band and conduction band. A large number of conduction electrons are available in them even at low temperature.

18. The Brillouin zone is a primitive unit cell of the reciprocal lattice. In mathematics and solid state physics, the first Brillouin zone is a uniquely defined primitive cell in reciprocal space.

Exercise

(A) Short Answer Type Questions :

1. What is the main assumption in classical free electron theory ?

2. What are the drawbacks of classical free electron theory ?

3. What is the main feature of Sommerfeld theory ?

4. What is the main drawback of Sommerfeld theory ?

5. Why Sommerfeld theory is unable to explain some solids are conductors, some are semiconductors and the others are insulators ?

6. Explain how Sommerfeld modified the classical free electron theory.

7. Write the Fermi-Dirac distribution function. Plot it as a function of energy.

8. What is Fermi level and Fermi energy ?

9. What do you understand by 'Fermi level' ? Plot this function for various temperatures including $T = 0°K$.

10. Write down the Schrödinger's equation for the electron in a one-dimensional potential box.

11. What are values of energy gap for metals, semiconductors and insulators ?

12. What is mobility?

13. On which factor Hall coefficient depends ?
14. How does Fermi energy change with temperature ?
15. State Bloch theorem.
16. Explain the basis of band theory.
17. Plot E versus K for nearly free electron model.
18. What do you mean by group velocity ?
19. What do you mean by density of states ?
20. Why the effective number of free electrons in completely filled band is zero ?
21. Why semiconductors are insulators at $0°K$ temperature ?
22. Why conductivity of semiconductor increases with increase in temperature ?
23. Why conductivity of metal decreases with increase in temperature ?

(B) Long Answer Type Questions :
1. Explain classical free electron theory.
2. Explain Wiedemann - Franz law.
3. Explain Sommerfeld's free electron theory and give the drawbacks of this theory.
4. Obtain an expression for energy levels and density of states in one dimension.
5. Describe Hall effect. Obtain an expression for Hall angle.
6. On the basis of band theory, distinguish between insulators, semiconductor and metals.
7. Explain Fermi energy and Fermi levels at $0°K$.
8. Define Fermi energy. Using Fermi distribution function, explain how does Fermi energy change with temperature.
9. Define Fermi energy. Show that $\bar{E} = \dfrac{3}{5} E_{F_0}$.
10. Define the term density of states. Obtain an expression for density of states in three dimensions.
11. Discuss the motion of a electron in cubical box and obtain an expression for its energy levels.
12. Explain the basis of band theory and describe the formation of energy bands in solid.
13. State the explain Bloch theorem.
14. Write a note on nearly free electron model.
15. On the basis of the nearly free electron model, show that at the Brillouin zone boundary energy gap exist.
16. Explain the origin of energy gap in a crystalline solid and calculate its magnitude.

17. Show that effective mass of an electron in a crystal is given by

$$m^* = \frac{\hbar^2}{\left(\dfrac{d^2E}{dK^2}\right)}$$

Interpret the result.

18. Distinguish between metals, semiconductors and insulators on the basis of band theory.

19. Obtain an expression for total effective number of free electrons in the given energy band. Discuss various cases.

20. What is Hall effect ? Show that Hall coefficient is equal to 1/ne.

(C) Unsolved Problems :

1. Calculate the energy difference between the $n_x = n_y = n_z = 1$ level and next higher level for free electrons in a solid of size 1 cm $\times$ 1 cm $\times$ 1 cm.　　**(Ans.** 1.1×10^{-14} eV)

2. The Fermi level of a solid is 5.51 eV. What is the average energy of the free electrons in the given solid at 0 K ?　　**(Ans.** 3.3 eV)

3. The Fermi energy of sodium is 3.15 eV. Assuming that it is the maximum kinetic energy of electrons in sodium, find the number of molecules per unit volume in sodium.　　**(Ans.** 2.543×10^{22} atoms/cm^3)

4. Calculate the Fermi energy of sodium from the given data :

   Density of sodium $= 1.01$ g/cm^3

   Atomic weight of sodium $= 23$

   Mass of the electron $= 9.1 \times 10^{-28}$ g

   Planck's constant, $h = 6.62 \times 10^{-27}$ ergs-sec.

   1 eV $= 1.602 \times 10^{-12}$ ergs.　　**(Ans.** $E_F = 3.234$ eV)

5. Calculate the energy difference between the $n_x = n_y = n_z = 1$ level and next higher energy level for free electrons in a solid cube of side 12 mm.

   Given : $h = 6.626 \times 10^{-34}$ joule-sec., $m = 9.1 \times 10^{-31}$ kg.

   Hint :
   $$E_1 = \frac{\hbar^2\pi^2}{2mL^2}\left(n_x^2 + n_y^2 + n_z^2\right) = \frac{3\pi^2\,\hbar^2}{2mL^2}$$

   $$E_2 = \frac{\hbar^2\pi^2}{2mL^2}\left[n_{x+1}^2 + n_x^2 + n_x^2\right] = \frac{6\pi^2\,\hbar^2}{2mL^2}$$

   $$E_2 - E_1 = \frac{\pi^2\,\hbar^2}{2mL^2}(6 - 3)$$

   $$E_2 - E_1 = \frac{3\pi^2\,\hbar^2}{2mL^2}$$
   　　(Ans. 7.85×10^{-15} eV)

6. Find the probability of occupancy of a state of energy (a) 0.05 eV above the Fermi energy, (b) 0.05 eV below the Fermi energy, (c) equal to the Fermi energy. Assume a temperature of 300 K.   **(Ans.** (a) 0.126, (b) 0.873, (c) 0.5)

7. What is the probability at 400 K that a state at the bottom of the conduction band is occupied in silicon ? Given that E_g = 1.1 eV. Assuming that the Fermi energy is to be at the middle of the gap between the conduction and valence bands,

$$E - E_F = \frac{1}{2} E_g$$

$$p(E) = \frac{1}{1 + \exp\left(\dfrac{E - E_F}{k_B T}\right)}, \quad E - E_F = E_g$$

$$\text{Factor } \frac{E_g}{2kT} = 15.942$$

$$p(E) = 8.4 \times 10^{-6}$$

Chapter 4...

Magnetism

Contents ...

Heike Kamerlingh Onnes

Heike Kamerlingh Onnes (21st September 1853 – 21 February 1926) was a Dutch physicist and Nobel laureate. He exploited the Hampson-Linde cycle to investigate how materials behave when cooled to nearly absolute zero and later to liquefy helium for the first time. His production of extreme cryogenic temperatures led to his discovery of superconductivity in 1911 for certain materials. Electrical resistance abruptly vanishes at very low temperatures.

Introduction

- After the discovery of electron it was known that the high electrical and thermal properties of the metals are attributed to the motion of electrons inside the metal. Also the magnetic properties of the materials are attributed to the electrons in the orbits of the atoms of the material. All magnetic effects in material can be explained on the basis of the current loops due to orbital and spin motion of the electrons. Due to these orbital and spin angular momentum of the electron, there is atomic magnetic moment associated with it. The intrinsic magnetic moment of the electron is the main source of magnetism in matter.

- Our description in magnetism in matter is based in part on the experimental facts that the presence of bulk matter generally modifies the magnetic field produced by these current loops. The material may have intrinsic magnetic moments or they may have induced magnetic moments due to external applied magnetic field on induction.

- On the basis of behavior on materials in the external magnetic field, the magnetic materials can be classified as – diamagnetic, paramagnetic and ferromagnetic. Diamagnetic materials are those whose atoms do not have permanent magnetic dipole moments. Paramagnetic and ferromagnetic materials are those which have atoms with permanent magnetic dipole moments. Diamagnetic materials are weakly repelled in the external magnetic field. Paramagnetic materials are those which are weakly attracted in external magnetic field and ferromagnetic materials are strongly attracted in the external field. In this chapter, we will discuss some microscopic properties of these materials.

4.1 Magnetization and Magnetic Field Strength   (Oct. 15)

- The magnetization of a substance is described by the magnetization vector, $\vec{M}$. **The magnitude of the magnetization vector is defined as the net magnetic dipole moment per unit volume**. The magnetic induction $\vec{B}$, which is the sum of external field and the field produced due to the current loops of electrons in the magnetic substance is related with magnetization by the relation,

$$\vec{B} = \mu_0 \vec{M} + \mu_0 \vec{H} \qquad \qquad ...(4.1)$$

  where μ_0 is permeability of free space and $\vec{H}$ is called **magnetic field strength**. Unit of B is Wb/m^2 and that of H and M is ampere per meter.

- In a large class of substances, specifically paramagnetic and diamagnetic substances, the magnetization vector $\vec{M}$ is proportional to $\vec{H}$. For these substances, we can write

$$\vec{M} = \chi \vec{H} \qquad \qquad ...(4.2)$$

  where the dimensionless quantity χ is called **magnetic susceptibility**. If the material is paramagnetic, χ is positive and the magnetization $\vec{M}$ is in the direction of $\vec{H}$. If the material is diamagnetic, χ is negative and the magnetization $\vec{M}$ is in the opposite direction of $\vec{H}$. It should be noted that this linear relationship does not apply for the ferromagnetic materials. If we use equation (4.2) in equation (4.1), we get,

$$\vec{B} = \mu_0(1 + \chi)\vec{H}$$

or $$\vec{B} = \mu_m\vec{H}$$

where **μ_m is called magnetic permeability** of the material and has value $\mu_m = \mu_0(1 + \chi)$. Substances may also be classified in terms of their magnetic permeabilities μ_m. We have,

For paramagnetic substances, $\mu_m > \mu_0$

For diamagnetic substances, $\mu_m < \mu_0$

For ferromagnetic substances, $\mu_m >> \mu_0$

4.2 Diamagnetism (Oct. 16)

- Diamagnetic substances are weakly repelled by in the external magnetic field and have negative magnetic susceptibility. Diamagnetism is small and has very weak effect in many materials caused by reactions of orbiting electrons to an applied magnetic field in accordance with Lenz law. In diamagnetic materials the magnetization is opposite to the direction of field of induction, so that χ is negative. The value of magnetic induction $\vec{B}$ is smaller in the diamagnetic material that it would be if the material is absent. Antimony Bismuth, mercury, gold and copper are examples of diamagnetic substances.

- If magnetic material is kept in magnetic fields, it can increase or decrease flux density. Diamagnetic materials reduce the density of lines of forces while paramagnetic substances increase flux density. Diamagnetism is created by circulation of electrons in atoms or molecule. For such diamagnetic materials, $\chi < 0$. It is independent of temperature.

- Diamagnetic materials have zero magnetic moments in the absence of external field. When external field is applied, net magnetic dipole moment opposing the field is induced in atom or molecule. Paramagnetic substances may have diamagnetic effect but it is much weaker compared with paramagnetic. Diamagnetic substance reduces the density of lines of forces while paramagnetic substances increase the flux density. The susceptibility of water is $- 9 \times 10^{-6}$ while that of graphite is $- 6 \times 10^{-5}$.

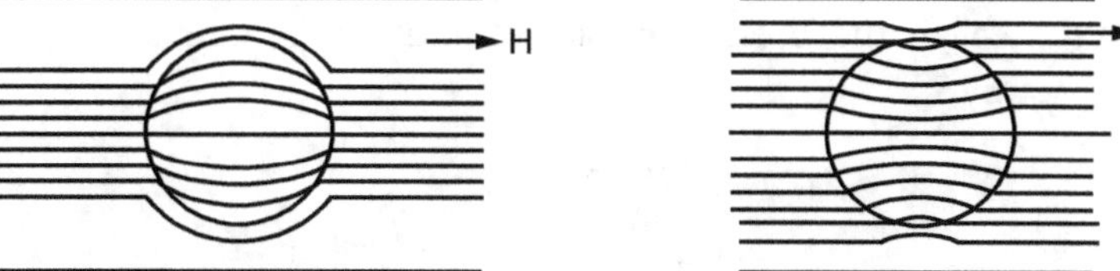

(a) Flux density in a diamagnetic sample   **(b) Flux density in a paramagnetic sample**

Fig. 4.1

4.3 Classical Theory of Diamagnetism (Langevin Theory)　(Oct. 16)

- Diamagnetism is associated with the tendency of electrical charges partially to shield the interior of a body from an applied magnetic field. In electromagnetism we are familiar with Lenz's law: when the flux through an electrical circuit is changed, an induced (diamagnetic) current is set up in such a direction as to oppose the flux change. The physical origin of diamagnetism can be seen from the classical picture of the atom.

- We will consider the simplest case, an electron revolving around the nucleus in a circular orbit of radius r and velocity v. The electron revolving in the circular orbit constitutes the current given by

$$I = \frac{e}{T} \qquad\qquad ...(4.3)$$

where T is the period of revolution of the electron. As v is the velocity of the electron in the circular orbit, the period of revolution is given by

$$T = \frac{2\pi r}{v} \qquad\qquad ...(4.4)$$

Substituting equation (4.4) in equation (4.3), we get

$$I = \frac{ev}{2\pi r} \qquad\qquad ...(4.5)$$

The magnitude of magnetic moment due to this current is given by

$$M = current \times area\ of\ the\ loop$$

$$\therefore \qquad M = I \times \pi r^2$$

$$or \qquad M = \frac{ev}{2\pi r} \times \pi r^2$$

$$\therefore \qquad M = \frac{evr}{2} \qquad\qquad ...(4.6)$$

Multiplying and dividing RHS of above equation by mass of electron m, we get

$$M = \frac{e}{2m}\ mvr$$

where mvr = l is called the orbital angular momentum. Therefore, magnetic moment vectorially is given by

$$\vec{M} = \frac{e}{2m}\ \vec{l}$$

Due to negative charge of electron, the direction of magnetic moment is opposite to that of orbital angular momentum as shown in Fig. 4.2.

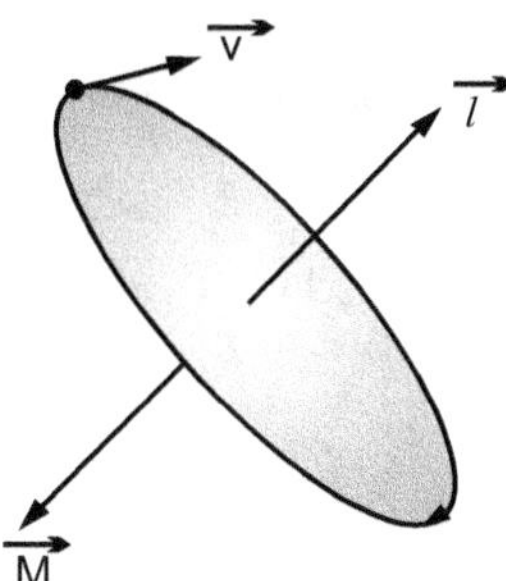

Fig. 4.2 : Direction of magnetic moment

- The origin of diamagnetism is Lenz's law. The atom with its circulating electron acts like an inductance in which counter e.m.f. is produced in accordance with Lenz's law. Since no resistance is encountered by the electron during its motion around the nucleus, the increased velocity does not diminish with time.

- Let us consider the effect of external applied magnetic field. Let B be the magnitude of external field which is normal to the orbit as shown in Fig. 4.3. The flux ϕ through the closed loop is

$$\phi = \pi r^2 B \qquad \qquad ...(4.7)$$

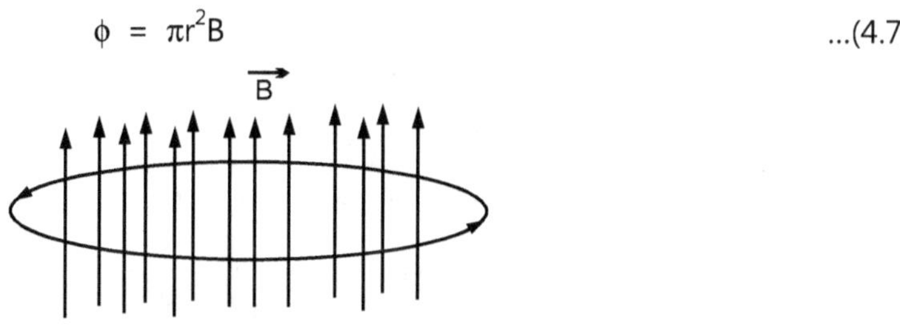

Fig. 4.3

If the field varies with time, the magnitude of the induced e.m.f. in the circuit is

$$\epsilon = -\frac{d\phi}{dt}$$

$$= -\pi r^2 \frac{dB}{dt}$$

This induced e.m.f. will act around the loop and set up an electric field E given by

$$\epsilon = 2\pi r E$$

$$\therefore \qquad 2\pi r E = -\pi r^2 \frac{dB}{dt}$$

or $$\qquad \vec{E} = -\frac{r}{2}\frac{d\vec{B}}{dt} \qquad \qquad ...(4.8)$$

This electric field exerts a force $e\vec{E}$ due to which the electron is accelerated. Therefore, we have

$$e\vec{E} = m\frac{d\vec{v}}{dt} \qquad \qquad ...(4.9)$$

$$\therefore \qquad \frac{d\vec{v}}{dt} = \frac{e\vec{E}}{m}$$

$$\text{or} \qquad \frac{d\vec{v}}{dt} = -\frac{er}{2m}\frac{d\vec{B}}{dt} \qquad \qquad ...(4.10)$$

Thus, in the time $\vec{B}$ changes by $\Delta\vec{B}$, the velocity $\vec{v}$ changes by $\Delta\vec{v}$. Therefore,

$$\Delta\vec{v} = -\frac{er}{2m}\Delta\vec{B} \qquad \qquad ...(4.11)$$

Using equation (4.6), we can write the induced magnetic moment as

$$\Delta\vec{M} = \frac{e\,r}{2}\Delta\vec{v}$$

$$\text{or} \qquad \Delta\vec{M} = -\frac{e^2 r^2}{4m^2}\Delta\vec{B} \qquad \qquad ...(4.12)$$

We know that $\vec{B} = \mu_0\vec{H}$, therefore, above equation can be written as

$$\Delta\vec{M} = -\frac{\mu_0 e^2 r^2}{4m^2}\Delta\vec{H} \qquad \qquad ...(4.13)$$

Magnetic susceptibility is given by equation (4.2). Using equation (4.2), equation (4.13) can be written as

$$\Delta\vec{M} = \chi\,\Delta\vec{H} \qquad \qquad ...(4.14)$$

where χ is given as

$$\chi = -\frac{\mu_0 e^2 r^2}{4m^2} \qquad \qquad ...(4.15)$$

If there are N atoms per unit volume, then the volume susceptibility can be written as

$$\chi = -\frac{N\mu_0 e^2}{4m^2}\sum r^2 \qquad \qquad ...(4.16)$$

The summation in above equation extends over all Z electrons in the atom, where Z is the atomic number. Since the core electrons have different radii, we can write

$$\sum r^2 = Z <r^2> \qquad \qquad ...(4.17)$$

where $< r^2 >$ is the mean square of the perpendicular distance from the field axis passing through the nucleus. If the axis is along Z-axis and electron orbit lies in xy plane, we have

$$< r^2 > \ = \ < x^2 > \ + \ < y^2 >$$

The mean square distance of the electron from the nucleus is

$$< R^2 > \ = < x^2 > + < y^2 > + < z^2 >.$$

For spherically symmetric charge distribution of charge, we have

$< x^2 > \ = \ < y^2 > \ = \ < z^2 >$. Therefore,

$$< r^2 > \ = \ \frac{2}{3} < R^2 > \qquad \qquad ...(4.18)$$

With equations (4.17) and (4.18), we can write equation (4.16) as

$$\chi \ = \ - \ \frac{\mu_0 N Z e^2}{6m} < R^2 > \qquad \qquad ...(4.19)$$

This is the classical **Langevin's formula for the susceptibility of the diamagnetic material**.

Following conclusions can be drawn from Langevin's formula :

1. All substances exhibit diamagnetism though it is masked by other magnetic effects.
2. The susceptibility of diamagnetic material is proportional to the atomic number.
3. If atom is bigger, the value of susceptibility is more.
4. For diamagnetic material, susceptibility is independent of temperature.

A perfect diamagnet, such as superconductor, excludes all the flux from the interior of the material so that $\vec{B}$ = 0 and χ = –1 for such materials.

4.4 Superconductors   (Oct. 17, 15)

- The electrical resistivity of many metals and alloys drops suddenly to zero when the specimen is cooled to a sufficiently low temperature, often a temperature in the liquid helium range. This phenomenon, called superconductivity, was observed first by Kamerlingh Onnes in 1911, three years after, he first liquefied helium. At a critical temperature T, the sample undergoes a phase transition from a state of normal electrical resistivity to a superconducting state.

- Thus superconductivity is a state of zero electrical resistance which occurs in certain materials below a characteristic temperature. The electrical resistivity of a metallic conductor decreases gradually as the temperature is lowered. However, in ordinary conductors such as copper and silver, this decrease is limited by impurities and other defects. Even near absolute zero, a real sample of copper shows some resistance. In metals such as copper and aluminium, electricity is conducted by outer energy level electrons they migrate individually from one atom to another. These atoms form a vibrating lattice within the metal conductor.

- When temperature is more, more is the vibrations of lattice and more is the resistance. As the electrons begin moving, they collide with tiny impurities or imperfections in the lattice. When the electrons bump into these obstacles they fly off in all directions and lose energy in the form of heat. Inside a superconductor the behavior of electrons is vastly different. The impurities and lattice are still there, but the movement of the superconducting electrons through the obstacle course is quite different. Despite these imperfections in a superconductor, the resistance drops abruptly to zero when the material is cooled below its critical temperature as shown in Fig. 4.4. An electric current flowing in a loop of superconducting wire can persist indefinitely with no power source.

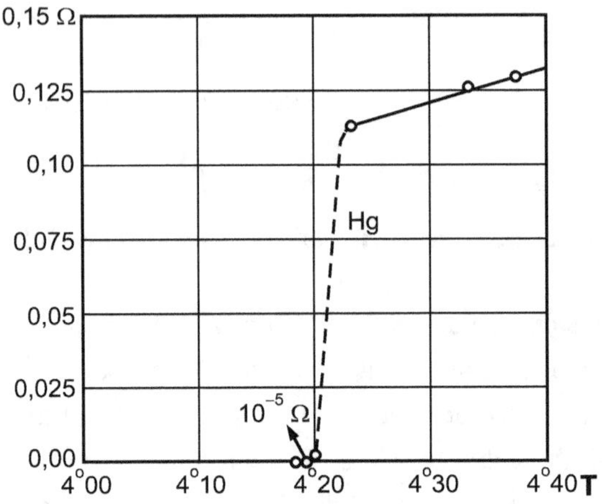

Fig. 4.4 : Resistance in ohms of a specimen of mercury versus absolute temperature

- Thus the first characteristic property of superconductor is its electrical resistance. For all practical purposes it is zero below a well defined temperature T_c called as critical or transition temperature.

4.5 Occurrence of Superconductivity　　　　　　(Oct. 17)

- The understanding of superconductivity was advanced in 1957 by three American physicists-John Bardeen, Leon Cooper and John Schrieffer, through their theories of superconductivity, known as the BCS Theory. The BCS theory explains superconductivity at temperatures close to absolute zero. Cooper realized that atomic lattice vibrations were directly responsible for unifying the entire current. They forced the electrons to pair up into teams. These pairs could pass all of the obstacles which caused resistance in the conductor. These teams of electrons are known as Cooper pairs. This pairing is caused by all attractive force between electrons from the exchange of phonons. Cooper and his colleagues knew that electrons which normally repel one another must have overwhelming attraction in superconductors. The electrons are ordered at temperatures below the transition temperature, and they are disordered above the transition temperature.

- The BCS theory successfully shows that electrons can be attracted to one another through interactions with the crystalline lattice. This occurs despite the fact that electrons have the same charge. When the atoms of the lattice oscillate as positive and negative regions, the electron pair is alternatively pulled together and pushed apart without a collision. The electron pairing is favorable because it has the effect of putting the material into a lower energy state. When electrons are linked together in pairs, they move through the superconductor in an orderly fashion. Individual electrons are fermions. In Cooper pairs they behave like bosons. The interaction between electrons in Cooper pair is by exchange of phonons.

- As long as the superconductor is cooled to very low temperatures, the Cooper pairs stay intact, due to the reduced molecular motion. As the superconductor gains heat energy, the vibrations in the lattice become more violent and break the pairs. As they break, superconductivity diminishes. Superconducting metals and alloys have characteristic transition temperatures (T_C) from normal conductors to superconductors. Below the superconducting transition temperature, the resistivity of a material is exactly zero. Superconductors made from different materials have different T_C values. Conventional superconductors usually have critical temperatures ranging from around 20 K to less than 1 K. Solid mercury, for example, has a critical temperature of 4.2 K. Lead (Pb) was found to superconduct at 7 K, and in niobium nitride (NbN) was found to superconduct at 16 K.

- The magnetic properties exhibited by superconductors are as dramatic as their electrical properties. It is observed experimentally that a bulk superconductor in a weak magnetic field act as a perfect diamagnet, with zero magnetic induction in the interior. When a superconductor specimen is placed in a magnetic field and cooled below transition temperature for superconductivity, the magnetic field originally present is ejected from the specimen as shown in Fig. 4.5. This is called as Meissner effect.

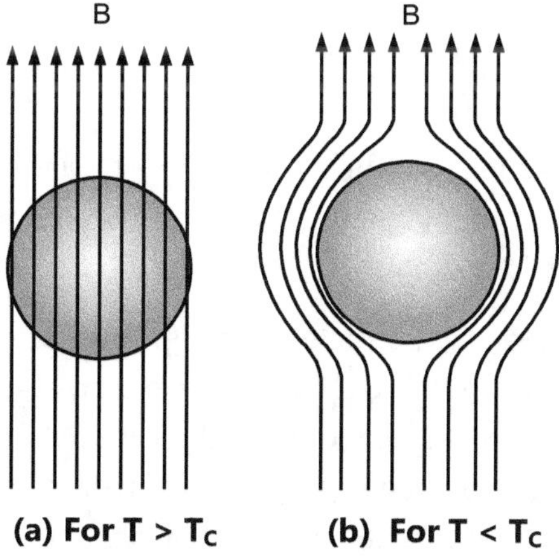

Fig. 4.5

4.6 Critical Magnetic Field and Meissner Effect   (April 17; Oct. 17, 15)

- The Meissner effect is the expulsion of a magnetic field from a superconductor during its transition to the superconducting state as shown in Fig. 4.5. Walther Meissner and Robert Ochsenfeld discovered the phenomenon in 1933 by measuring the magnetic field distribution outside the superconducting tin and lead samples. The samples in the presence of an applied magnetic field were cooled below their superconducting transition temperature. Below the transition temperature the samples cancelled all magnetic fields inside, which mean they became perfectly diamagnetic. They detected this effect only indirectly; because the magnetic flux is conserved by a superconductor, when the interior field decreased the exterior field increased. The experiment demonstrated for the first time that superconductors were more than just perfect conductors and provided a uniquely defining property of the superconducting state.

- A superconductor with little or no magnetic field within it is said to be in the Meissner state. Superconductors in the Meissner state exhibit perfect diamagnetism, or superdiamagnetism, meaning that the total magnetic field $\vec{B} = 0$ within them. It shows that a bulk superconductor behaves in an applied external field B_a as if inside the specimen $B = 0$. The magnetic induction B, which is the sum of external field and the field of magnetization is zero i.e.

$$\vec{B} = \vec{B_a} + \mu_0 \vec{M} = 0$$

$$\vec{B} = \mu_0 \vec{H_c} + \mu_0 \vec{M} = 0$$

where H_c is the magnetic intensity of applied field.

$$\therefore \qquad \frac{M}{B_a} = -\frac{1}{\mu_0} \quad \text{or} \quad \frac{M}{H_c} = -1.$$

- This means that their magnetic susceptibility, $\chi_m = -1$. Diamagnetics are defined by the generation of a spontaneous magnetization of a material which directly opposes the direction of an applied field.

- When the temperature is lowered to below critical temperature, (T_C), the superconductor will "push" the field out of itself. It does this by creating surface currents in itself which produces a magnetic field exactly countering the external field, producing a "magnetic mirror". The superconductor becomes perfectly diamagnetic, cancelling all magnetic flux in its interior. This perfect diamagnetic property of superconductors is perhaps the most fundamental macroscopic property of a superconductor.

- The magnetization versus applied field curve expected for a superconductor under the conditions of Meissner-Ochsenfeld experiment is shown in Fig. 4.6. Pure specimens of many materials exhibit this behaviour, they are called **Type-I superconductors**. Very pure samples of lead, mercury, and tin are examples of Type-I superconductors. In Type-I superconductors, superconductivity is abruptly destroyed when the strength of the applied field rises above a critical value H_c.

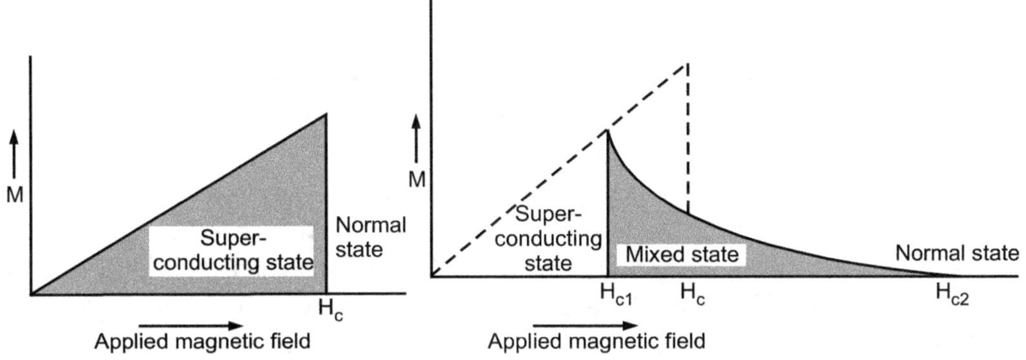

Fig. 4.6 : Type-I superconductors   **Fig. 4.7 : Type-II superconductors**

- **A Type-II superconductor** is a superconductor characterized by its gradual transition from the superconducting to the normal state within an increasing magnetic field (Refer Fig. 4.7). Typically they superconduct at higher temperatures and magnetic fields than Type-I superconductors. This allows them to conduct higher currents and makes them useful for strong electromagnets.

- In comparison to the (theoretically) sharp transition of a Type-I superconductor above the lower temperature T_{c1}, magnetic flux from external field is no longer completely expelled, and the superconductor exists in a mixed state. Above the higher temperature T_{c2}, the superconductivity is completely destroyed and the material exists in a normal state. Both of these temperatures are dependent on the strength of the applied field.

- It is more usual to consider a fixed temperature, in which case transition (flux penetration) occurs between critical field strengths H_{c1} and H_{c2} (the upper critical field). Between the lower critical field H_{c1} and the upper critical field H_{c2}, the flux $B \neq 0$ and the Meissner effect is incomplete. In the region between H_{c1} and H_{c2} the magnetic flux lines run through narrow regions of non-superconducting material, surrounded by vortices of supercurrents protecting the rest of the superconductor. This state is said to be **vortex state**.

- Niobium, Vanadium, Technetium, Diamond and Silicon are pure element. Type-II superconductors are usually made of metal alloys like niobium-titanium, niobium-tin also show Type-II behaviour. They are generally high temperature superconductors. Other examples of Type-II superconductors are the cuprate-perovskite ceramic materials which have achieved the highest temperatures to reach the superconducting state for example, Yttrium barium copper oxide ($YBa_2Cu_3O_7$) and Bismuth strontium calcium copper oxide ($Bi_2CaSr_2Cu_2O_9$).

Critical Magnetic Field

- The Meissner state breaks down when the applied magnetic field is too large. At a fixed temperature below the critical temperature, superconducting materials cease to superconduct when all external magnetic field is applied which is greater than the *critical magnetic field*. The superconducting state cannot exist in the presence of a magnetic field greater than a critical value, even at absolute zero. This critical magnetic field is strongly correlated with the critical temperature for the superconductor, which is in turn correlated with the band gap. Type-II superconductors show two critical magnetic field values, one at the onset of a mixed superconducting and normal state and one where superconductivity ceases (Refer Fig. 4.8).

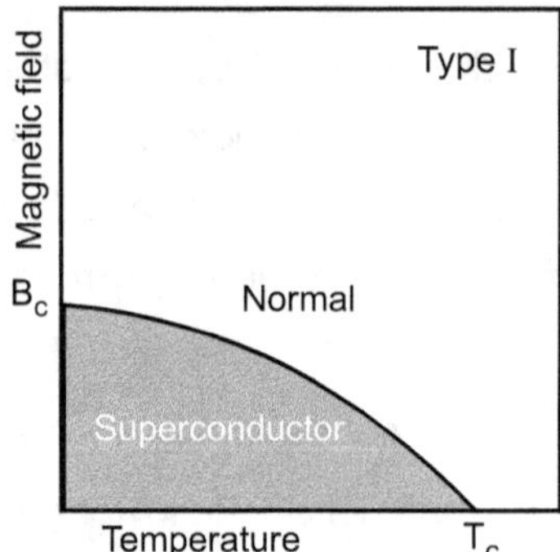

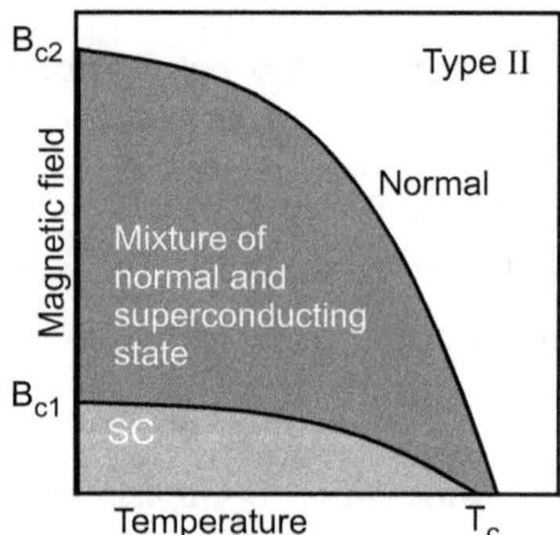

Fig. 4.8 : Critical magnetic field

- It is the nature of superconductors to exclude magnetic fields (Meissner effect) so long as the applied field does not exceed their critical magnetic field. This critical magnetic field is tabulated for 0 K and decreases from that magnitude with increasing temperature, reaching zero at the critical temperature for superconductivity. The critical magnetic field at any temperature below the critical temperature is given by the relationship

$$B_c = B_c(0)\left[1 - \left(\frac{T}{T_c}\right)^2\right]$$

APPLICATIONS OF SUPERCONDUCTIVITY   **(April 16)**

The potential applications of superconductivity in various areas are listed as in Table 4.1.

Table 4.1 : Applications of superconductivity

Area	Applications
Magnet	• High field energy magnets • NMR • Magnetic levitation • Magnetic shielding • Large physical machines (collider, fusion confinement)
Energy	• Production by magnetic fusion and magneto hydrodynamics • Magnetic energy storage • Electric power transmission
Transportation	• High speed train • Ship drive system
Electronics	• SQUID (Superconducting quantum interface devices) • Bolometer • Electromagnetic shielding • Josephson devices (square law detector, paramagnetic amplifier, mixer)
Computer and IT	• Optoelectronics • Voltage standard • Active superconducting materials • Semiconductors

• Superconductors are being used in many research and development field. Superconducting magnets are already taking precedence over conventional magnets in many types of equipment. Electrical machines, transformers and cables are being developed using superconductors.

4.7 Paramagnetism   (April 17, 16)

• In a paramagnetic substance each atom contains permanent magnetic dipole moment. Due to thermal agitation, these dipole moments are randomly oriented in the material giving zero net magnetic moment in the absence of external field. These moments are due to the intrinsic electronic spin and the orbital motion of the electrons.

- If an external magnetic field $\vec{B}$ is applied, the magnetic dipole moments of the atoms are oriented along the field direction. This is because the energy is lower when the magnetic dipole moment is parallel to the field than when it is antiparallel, so the parallel alignment is preferred. Due to this the induced field in the material adds to the applied field so that susceptibility is positive. Though there are some diamagnetic effects, they are negligible.

- The tendency of the magnetic dipole moments to align along the applied field direction is opposed by the thermal motion which tends to make the directions of the magnetic dipole moments random. Hence the susceptibility is temperature dependent. In atoms with filled subshells, the spin magnetic moments and orbital magnetic moments separately cancel in pairs. Paramagnetism is shown by the atoms which have unpaired electrons *i.e.* electronic subshells are partly filled.

- Let us now calculate the paramagnetic susceptibility for the simplest kind of system, that is one containing separated atoms. The magnetic moment of the atom in free space is given by

$$\vec{\mu} = -g\,\mu_B\,\vec{J} \qquad\qquad ...(4.20)$$

where μ_B is called Bohr magneton and g is the Lande's g factor. For a free electron, g is given by

$$g = 1 + \frac{J(J+1) + L(L+1) + S(S+1)}{2J(J+1)}$$

- The Bohr magneton in SI unit is defined as $e\hbar/2m$. Its numerical value is $\mu_B = 9.274 \times 10^{-24}$ A.m^2 $= 9.274 \times 10^{-24}$ J/T. The magnitude of the magnetic dipole moment associated with a given value of J is given by

$$\mu = g\mu_B\sqrt{J(J+1)}$$

- Let an external magnetic field $\vec{B}$ be applied to the material. The energy levels of the system in the magnetic field are

$$U = -\vec{\mu}\cdot\vec{B}$$

$$= m_J g\mu_B B \qquad\qquad ...(4.21)$$

where m_J is the azimuthal quantum number and has the values $J, J-1, J-2, ..., 0, ... -J+1, J$. For a single spin with no orbital motion, we have $m_J = \pm1/2$ and $g = 2$, hence we have $U = \pm\mu_B B$. This splitting is shown in Fig. 4.9.

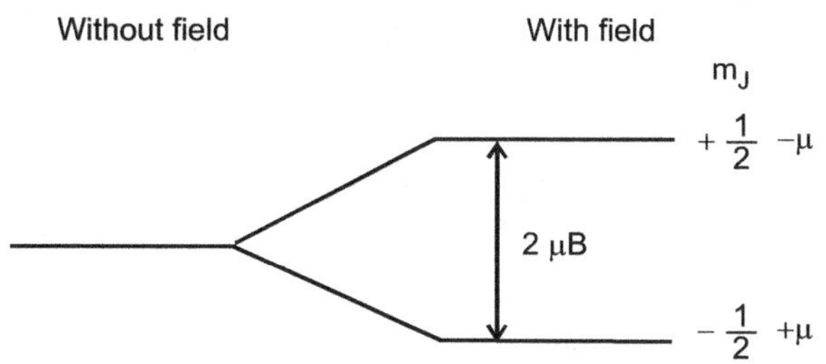

Fig. 4.9

4.8 Langevin Theory of Paramagnetism   (April 17, 16)

- Consider a system containing separated atoms, in each of which the electrons orbital angular momentum is zero and there is unpaired electron with two space orientations of spin. Let N be the number of unpaired magnetic dipole moments of electrons per unit volume of the material. If N_- represents the volume density of moments that are parallel to the field and N_+ represents the same for the moments antiparallel to the field, then $N = N_- + N_+$. The magnetic potential energy for electron with parallel orientation with field direction is $-\mu B$ and that of antiparallel orientation is μB. Then, from the Boltzmann distribution law, the number of electrons in each energy state that is parallel to the field as

$$N_- = AN\,e^{-\mu B/kT}$$

where A is some proportionality constant and k is Boltzmann constant.

Similarly, we have $\qquad N_+ = AN\,e^{\mu B/kT}$

The resultant magnetization i.e. magnetic dipole moment per unit volume is given by

$$M = \mu\,(N_- - N_+)$$

or $\qquad M = \mu AN(e^{-\mu B/kT} - e^{\mu B/kT})$

The average net magnetic moment is given by

$$\bar{\mu} = \frac{M}{N} = \frac{\mu AN(e^{-\mu B/kT} - e^{\mu B/kT})}{(N_- + N_+)}$$

or $\qquad \bar{\mu} = \mu\,\dfrac{(e^{-\mu B/kT} - e^{\mu B/kT})}{(e^{-\mu B/kT} + e^{\mu B/kT})} \qquad\qquad \ldots(4.22)$

The net magnetic dipole moment per unit volume i.e. magnetization is given by

$$M = N\bar{\mu}$$

$$M = N\mu\,\frac{(e^{-\mu B/kT} - e^{\mu B/kT})}{(e^{-\mu B/kT} + e^{\mu B/kT})} \qquad\qquad \ldots(4.23)$$

Since under ordinary circumstances, $\mu B \ll kT$ and we can express $e^{\mu B/kT} = 1 + \mu B/kT + \ldots$ and $e^{-\mu B/kT} = 1 - \mu B/kT + \ldots$. Using these in equation (4.22), we get

$$M = N\mu \frac{(1 - \mu B/kT) - (1 + \mu B/kT)}{(1 - \mu B/kT) + (1 + \mu B/kT)}$$

$$\therefore \qquad M = \frac{N\mu^2 B}{kT}$$

Since $B = \mu_0 H$, we can write the above equation as

$$M = \frac{N\mu^2 \mu_0 H}{kT} \qquad\qquad \ldots(4.24)$$

The magnetic susceptibility is given by

$$\chi = \frac{M}{H}$$

Therefore, from equation (4.24), we get

$$\chi = \frac{N\mu^2 \mu_0}{kT} \qquad\qquad \ldots(4.25)$$

This is **Langevin result for paramagnetic susceptibility**.

The equation (4.25) can be written as

$$\chi = \frac{C}{T} \qquad\qquad \ldots(4.26)$$

This equation is called Curie equation and the constant $C = N\mu^2 \mu_0 / T$ is called **Curie constant**. The equation (4.26) shows that the susceptibility varies inversely as temperature.

Fig. 4.10 shows the plot of magnetization M from equation (4.23) against the applied field B. It is observed that M approaches the value $N\mu$ asymptotically.

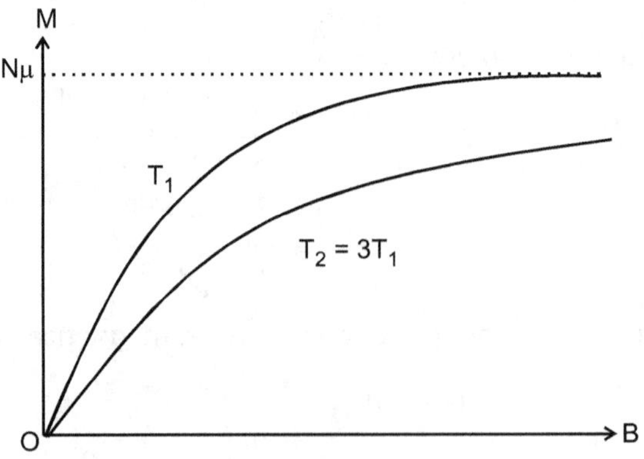

Fig. 4.10

4.9 Ferromagnetism

- When we talk about a magnet, we are referring to ferromagnet such as a piece of iron which has the ability to pick up other pieces of iron. Such material is called as permanent magnet as it exhibits magnetic properties even in absence of applied field. However it has been observed that magnetization increases when external magnetic field is applied to the specimen. Ferromagnetic materials exhibit hysteresis loop.

- The ferromagnetism in Fe, Co, Ni and Gd is mainly due to self alignment of groups or atoms carrying a permanent magnetic moments in the same direction. A ferromagnet has a spontaneous magnetic moments - a magnetic moment even in zero applied magnetic field. The existence of a spontaneous moment suggests that electron spins and magnetic moments are arranged in a regular manner. The order need not be simple: all of the spin arrangements sketched in Fig. 4.11 except the simple antiferromagnet have a spontaneous magnetic moment, called the **saturation moment.**

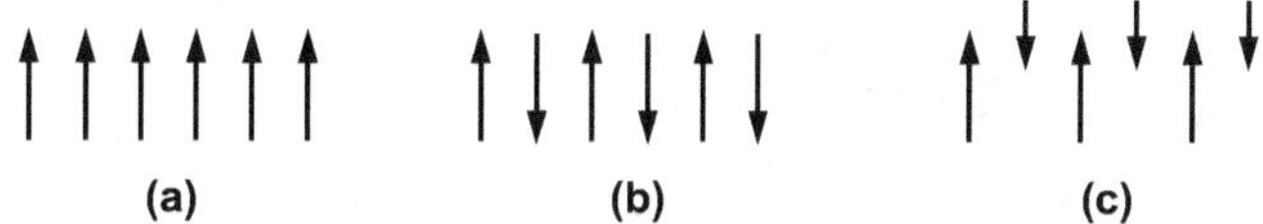

(a)   (b)   (c)

Fig. 4.11 : (a) Simple ferromagnet, (b) Simple antiferromagnet (c) Ferrimagnet

- For a given ferromagnetic material, the spontaneous magnetization can occur below a certain temperature T_c called ferromagnetic curie temperature. Above curie temperature, ferromagnetic materials behave like paramagnetic materials and have susceptibility given by Curie Weiss law.

$$\chi = \frac{C}{T - T_c}$$

- Above the transition temperature, ferromagnetic substances become paramagnetic, because then the parallel spin aligning energy is sufficient to overcome the randomizing effect. Table 4.2 indicates the curie temperature of major ferromagnetic materials.

Table 4.2

Element	Electronic configuration	Crystal structure	Magnetisation at 0 K (amp/m)	T_c (K)
Fe	$3d^6\,4s^2$	bcc	1.7×10^6	1043
Co	$3d^7\,4s^2$	hcp	1.4×10^6	1404
Ni	$3d^8\,4s^2$	fcc	0.48×10^6	632
Gd	$4f^7 5d^1\,6s^2$	hcp	5.66×10^7	290

- Above table indicates that ferromagnetism is not exclusively characteristic of crystal structure. They have wide range of curie temperature and a noticeable range of magnetic moment. One common thing is electronic structure. Three of them are found in 3d transition group and other in the 4f rare earth transition group. The existence of partly filled d or f shell is essential.

4.10 Ferromagnetic Domains　　　　　(April 16, Oct. 16)

- Ferromagnetism occurs when paramagnetic ions in a solid, 'lock' together in a small region in which all magnetic moments are along the same direction. Such a region is called **domain**. There are number of domains present in the ferromagnetic material. Each of these domains has a volume about 10^{-12} to 10^{-8} m^3 and contains 10^{17} to 10^{21} atoms. The boundaries between domains having different orientations are called **domain walls**.

- If there is no external magnetic field applied to the sample of ferromagnetic material, the domains are randomly oriented such that the net magnetic moment is zero, as shown in Fig. 4.12 (a). When the sample is placed in an external magnetic field, the domains tend to align along the field direction giving net magnetization, as shown in Fig. 4.12 (b).

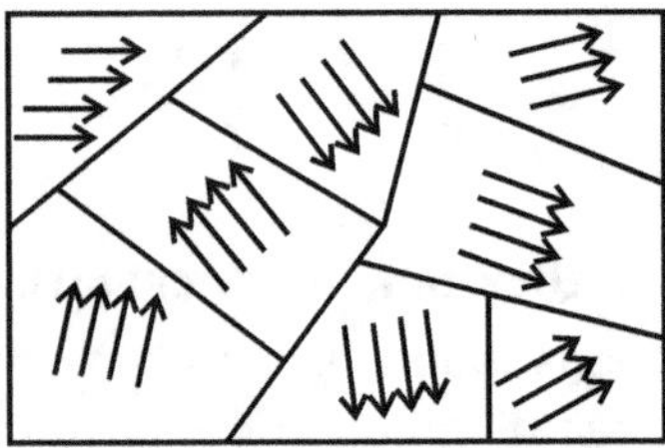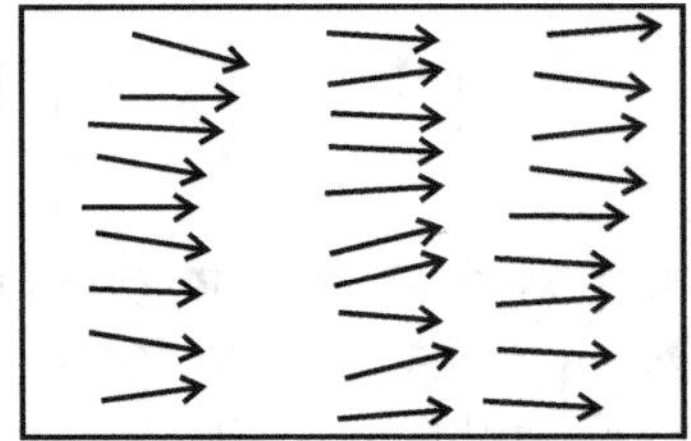

(a) Without external field　　　　　**(b) With external field**

Fig. 4.12 : Orientation of domains

- Experimental observations show that when an external field is applied, those domains that are aligned with the field direction grow in size at the expense of the less favorably oriented domains. When external field is very strong, all the domains are aligned along the field direction and provide the high observed magnetization. When the external field is removed, the sample may retain a net magnetization in the direction of original field. At ordinary temperatures, thermal agitation is not sufficient to disrupt this alignment to random alignment.

- If the temperature of the ferromagnetic substance is increased, at sufficiently high temperature the domain structure breaks down and ferromagnetic substance becomes paramagnetic. This temperature is called Curie temperature T_c.

- The number of ferromagnetic substances is limited. Iron, Cobalt and Nickel (all transition group elements) are ferromagnetic at sufficiently low temperatures. Rare earths such as gadolinium (Gd) and terbium (Tb) are ferromagnetic below room temperature, while other rare earths are ferromagnetic at very low temperatures. At sufficiently high temperatures, all transitions and rare earth metals become paramagnetic.

- For ferromagnetic substance the spontaneous magnetization is maximum at $T = 0°K$ and drops to zero at a temperature T_c, as shown in Fig. 4.13. T_c is called Curie temperature.

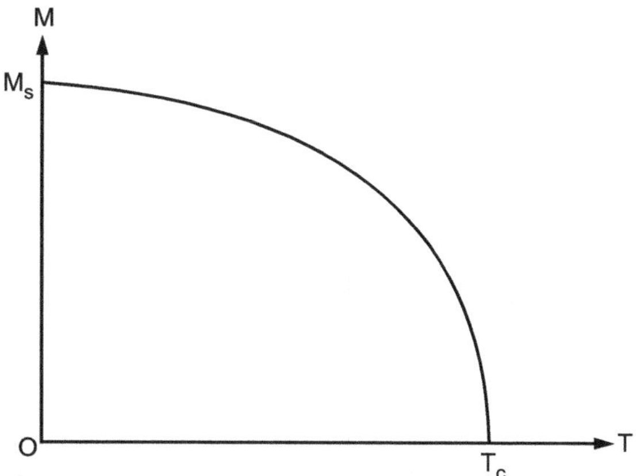

Fig. 4.13 : Plot of spontaneous magnetization against temperature

4.11 Hysteresis   (Oct. 16)

- A large amount of information can be obtained about the magnetic properties of a ferromagnetic material by studying hysteresis loop. Hysteresis comes Greek word (hysteros means later) and refers to the retardation in the response of material to a change in the applied field. It shows relationship between the induced magnetic induction ($\vec{B}$) and the magnetizing intensity ($\vec{H}$). A typical hysteresis loop i.e. B-H curve is shown in Fig. 4.14.

- The loop is generated by measuring the magnetic flux $\vec{B}$ while magnetizing force is changed. When fresh ferromagnetic material is magnetized, it will be magnetized along the dotted curve shown in Fig. 4.14. At point 'a' almost all of the magnetic domains are aligned and an additional increase in the magnetizing force will produce very little increase in magnetic flux. The material has reached the point of magnetic saturation. When magnetic intensity H is reduced to zero, the material is demagnetized along the path 'ab'. When H = 0, there is some magnetization remained in the material shown by point 'b'. This point is referred as **retentivity** and indicates the remanence or level of residual magnetism in the material. As the magnetic intensity $\vec{H}$ is reversed, at some negative 'H', the flux in the material becomes zero. This point is point 'c' in Fig. 4.14. This is called the point of **coercivity** on the curve. The magnetic force required to remove the residual magnetism from the material is called coercive force or coercivity of the material.

- As the magnetizing intensity H is further increased in the negative direction, the material will again become magnetically saturated but in the opposite direction (point 'd' on the curve). Reducing $\vec{H}$ to zero brings the magnetization to point 'e' on the curve. Increasing

$\vec{H}$ in positive direction will decrease the magnetic flux $\vec{B}$ = 0 at point 'f'. Again increase in $\vec{H}$ along positive direction, the material is again saturated at point 'a'. The material will never return to point O.

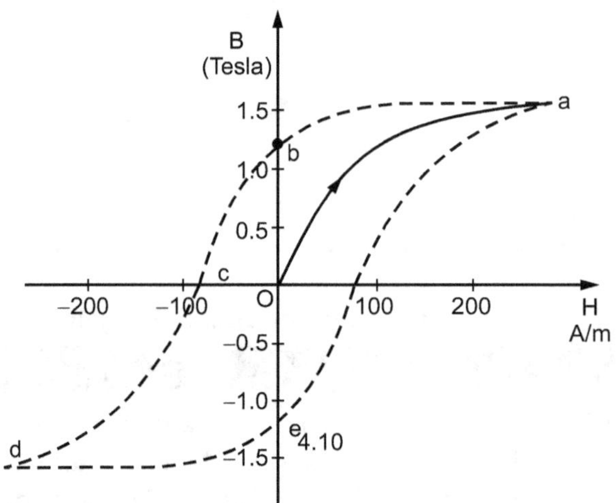

Fig. 4.14 : Hysteresis loop

- The value of the coercivity ranges over seven orders of magnitude; it is the most sensitive property of ferromagnetic materials which is subject to control. The coercivity may vary from 600 G in a loudspeaker permanent magnet (Alnico V) and 10,000 G in a special high stability magnet ($SmCo_5$) to 0.5 G in a commercial power transformer (Fe-Si 4 wt. pet.) and 0.002 G in a pulse transformer (Supermalloy). Low coercivity is desired in a transformer, for this means low hysteresis loss per cycle of operation. Materials with low coercivity are called soft and those with high coercivity are called hard. The coercivity decreases as the impurity content decreases and also as internal strains are removed by annealing (slow cooling). Amorphous ferromagnetic alloys may have low coercivity, low hysteresis losses, and high permeability.

4.12 Antiferromagnetism

- There are number of materials that show interesting modifications of ferromagnetic behavior. These materials are characterized by containing two interlocking sets of atoms or atom groupings each of which has the spin lining up character of ordinary ferromagnetic substances. In some of these materials, the two sublattices, as they are called, tend to magnetize in the same direction, and to behave like ferromagnetic substancesas shown in Fig. 4.15 (a). In some of the materials like MnO, the two sub lattices have identical magnetic moments but opposite in direction as shown in

Fig. 4.15 (b). These orientations are due to exchange interaction forces between adjacent atoms to have antiparallel spin orientations. Hence such materials show very little gross external magnetism. Such materials are called antiferromagnetism. If they are heated sufficiently, the materials become paramagnetic and the interactions ceasing to act.

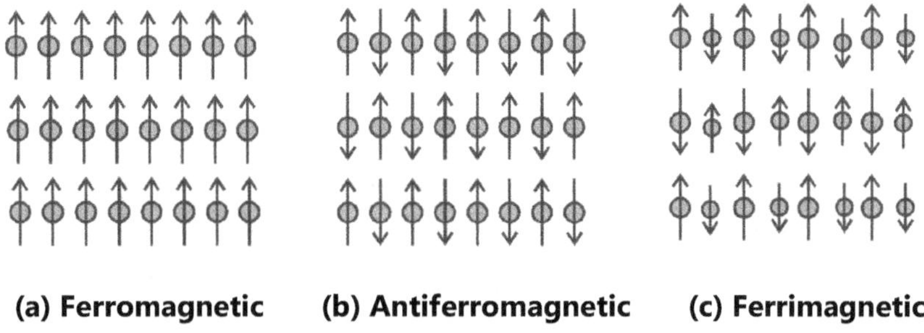

    (a) Ferromagnetic   **(b) Antiferromagnetic**   **(c) Ferrimagnetic**

Fig. 4.15

4.13 Ferrimagnetism   (Oct. 15)

- In materials in which the two sub-lattices are not identical, there will be a net effect similar to ferromagnetism but considerably weaker. The name of this phenomenon is **ferrimagnetism** and substances possessing it are called **ferrites**. For example, in nickel ferrite the two ions are Ni^{++} and Fe^{+++}. The exchange interaction locks the ions into a pattern like that of Fig. 4.15 (c).

- In ferromagnetic materials the same antiferromagnetic exchange interactions exists, which align the magnetic moments antiparallel. But since the ions are of different magnitudes of magnetic dipole moments, the net magnetization is not zero. In ferrimagnetic materials also the exchange interaction disappears if the material is heated above a certain characteristic temperature. The ferrites are crystals which have small electrical conductivity as compared to ferromagnetic materials.

4.14 Neel Temperature   (Oct. 17, April 17)

- The Neel temperature, T_N, is the temperature above which an antiferromagnetic material becomes paramagnetic, that is, the thermal energy becomes large enough to destroy the macroscopic ordering within the material. It is also called magnetic ordering temperature.

- The Neel temperature is analogous to the Curie temperature, T_C, for ferromagnetic materials. It is named after Louis Neel (1904-2000), who received the 1970 Nobel prize in physics for his work in the area.

- Table 4.3 indicates the Neel temperatures of several antiferromagnetic materials.

Table 4.3

Substance	Neel temperature, in K
MnO	122
MnS	165
MnTe	307
MnF_2	67
FeF_2	79
$FeCl_2$	24
FeO	198
$CoCl_2$	25
CoO	292

4.15 Ferrites and their Applications   (Oct. 17, 15; April 16)

- Ferrites are a class of ferrimagnetic ceramic chemical compounds consisting of mixtures of various metal oxides, usually including iron oxides. Their general chemical formula may be written as AB_2O_4, where A and B represent different metal cations. For example, $ZnFe_2O_4$, $CoFe_2O_4$, and $MgAl_2O_4$. Ferrites are usually non-conductive ferrimagnetic ceramic compounds derived from iron oxides such as hematite (Fe_2O_3) or magnetite (Fe_3O_4) as well as oxides of other metals. In ferrites the saturation magnetization at $T = 0$ K does not correspond to parallel alignment of the magnetic moments of the constituent ions, even in crystals for which there is strong evidence that the individual paramagnetic ions have their normal magnetic moments. Ferrites are, like most other ceramics, hard and brittle. In terms of the magnetic properties, ferrites are often classified as "soft" and "hard" which refers to their low or high coercivity of their magnetization, respectively.

Soft ferrites :

- The ferrites which have low coercivity are called **soft ferrites**. These ferrites that are used in transformer or electromagnetic cores contain nickel, zinc, and/or manganese compounds. The low coercivity means the material's magnetization can easily reverse direction without dissipating much energy (hysteresis losses), while the material's high resistivity prevents eddy currents in the core, another source of energy loss. Because of their comparatively low losses at high frequencies, they are extensively used in the cores of RF transformers and inductors in applications such as switched-mode power supplies (SMPS).

- The most common soft ferrites are manganese-zinc (MnZn, with the formula $Mn_aZn_{(1-a)}Fe_2O_4$) and nickel-zinc (NiZn, with the formula $Ni_aZn_{(1-a)}Fe_2O_4$). NiZn ferrites exhibit higher resistivity than MnZn and are therefore more suitable for frequencies above 1 MHz. MnZn have higher permeability and saturation induction than NiZn.

Hard ferrites :

- The ferrites which have a high coercivity and high remanence after magnetization are called hard ferrites. Permanent ferrite magnets are made of hard ferrites. These are composed of iron and barium or strontium oxides. In a magnetically saturated state they conduct magnetic flux well and have a high magnetic permeability. This enables these so-called **ceramic magnets** to store stronger magnetic fields than iron itself. They are cheap, and are widely used in household products such as refrigerator magnets. The maximum magnetic field B is about 0.35 tesla and the magnetic field strength H is about 30 to 160 kiloampere turns per meter (400 to 2000 oersteds). The density of ferrite magnets is about 5 g/cm^3.

Crystal structure of Ferrites :

- The cubic ferrites have spinel crystal structure. Ferrites adopt a crystal motif consisting of cubic close-packed (FCC) oxides (O^{2-}) with A cations occupying one-eighth of the octahedral holes and B cations occupying half of the octahedral holes [Refer Fig. 4.16 (a) and (b)]. The magnetic material known as "ZnFe" has the formula $ZnFe_2O_4$, with Fe^{3+} occupying the octahedral sites and half of the tetrahedral sites. The remaining tetrahedral sites in this spinel are occupied by Zn^{2+}.

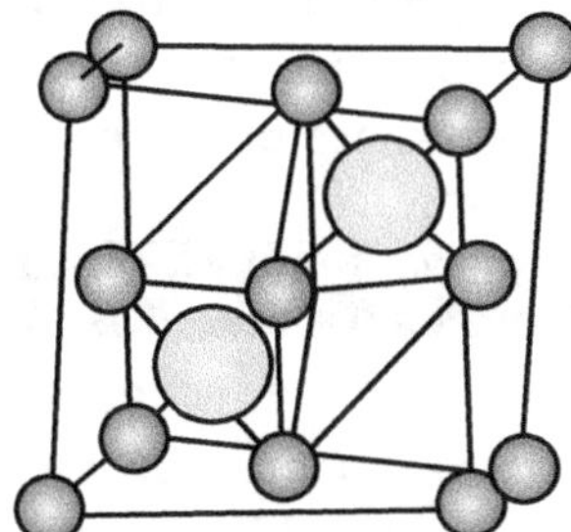

(a) For A cation

(Big sphere A and small sphere is O)

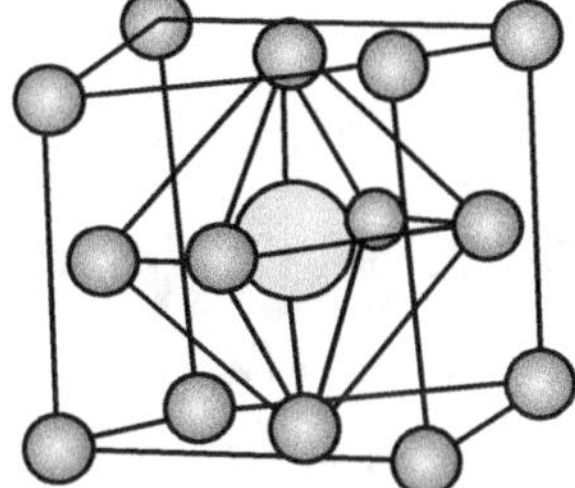

(b) For B cation

(Big sphere B and small sphere is O)

Fig. 4.16

Fig. 4.17 shows crystal structure of mineral spinel $MgAl_2O_4$.

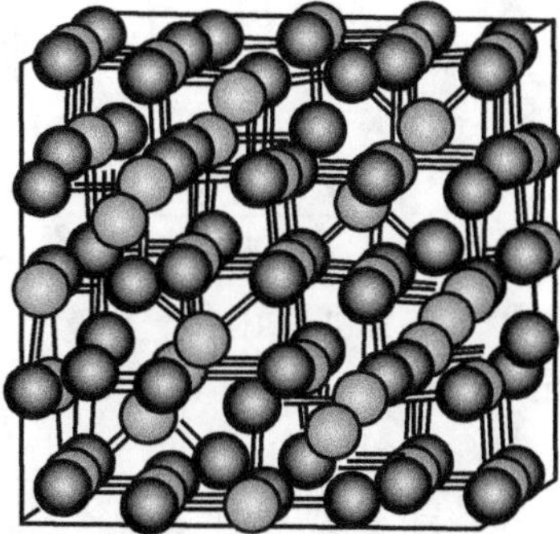

Fig. 4.17 : Structure of mineral spinel $MgAl_2O_4$

Applications of Ferrites :

1. Ferrite cores are used in electronic inductors, transformers, and electromagnets. The high electrical resistance of the ferrite leads to very low eddy current losses in these devices.

2. Ferrites are commonly seen as a lump in a computer cable, called a ferrite bead, which helps to prevent high frequency electrical noise (radio frequency interference) from exiting or entering the equipment. Ferrites are also used in broadband transformers.

3. Early computer memories stored data in the residual magnetic fields of hard ferrite cores, which were assembled into arrays of core memory.

4. Ferrite powders are used in the coatings of magnetic recording tapes of tape recorders.

5. Ferrite particles are also used as a component of radar-absorbing materials or coatings used in stealth aircraft and in the absorption tiles lining the rooms used for electromagnetic compatibility measurements.

6. Ferrite magnets are used in loudspeakers.

7. It is a common magnetic material for electromagnetic instrument pickups, because of price and relatively high output. However, such pickups lack certain sonic qualities found in other pickups such as those that use Alnico alloys or more sophisticated magnets.

Solved Examples

Example 4.1 : *A paramagnetic substance has 10^{28} atoms/m^3. The magnetic moment of each atom is 1.79×10^{-23} A-m^2. Calculate the paramagnetic susceptibility of the material at temperature 320 °K. What would be the dipole moment of the rod of this material 0.1 m long and 1 cm^2 cross-section placed in a field of 7×10^4 A/m ?*

Solution : The susceptibility of paramagnetic material is given by

$$\chi = \frac{N\mu^2\mu_0}{kT}$$

where $N = 10^{28}$ atoms/m^3, $\mu = 1.79 \times 10^{-23}$ A-m^2, $\mu_0 = 4\pi \times 10^{-7}$ Wb/A-m,

$k = 1.38 \times 10^{-23}$ J/°K, $T = 320$ °K, $H = 7 \times 10^4$ A/m.

Thus,
$$\chi = \frac{10^{28} \times (1.79 \times 10^{-23})^2 \times 4\pi \times 10^{-7}}{1.38 \times 10^{-23} \times 320}$$

$$= 9.11 \times 10^{-4}$$

The magnetization is given as

$$M = \chi H$$
$$= 9.11 \times 10^{-4} \times 7 \times 10^4$$
$$= \mathbf{63.77 \ A/m} \qquad\qquad \textbf{... Ans.}$$

The magnetization is given as net dipole moment per unit volume, therefore, magnetic dipole moment is

$$\mu = M \times V$$

where V = volume of the rod $= 0.1 \times 10^{-4} \, m^3 = 10^{-5} \, m^3$

$$\therefore \quad \mu = 63.77 \times 10^{-5} \, A\text{-}m^2 \qquad \text{... Ans.}$$

Example 4.2 : *Calculate the magnetization of 1 gm of oxygen gas at normal temperature and pressure in the earth's magnetic field. The susceptibility of the oxygen is 2.1×10^{-26} and earth's magnetic field is 5×10^{-5} tesla. Also calculate the dipole moment of each atom.*

Solution : The magnetization is given as

$$M = \chi H$$

We have

$$B = \mu_0 H$$

$$\therefore \quad M = \chi \frac{B}{\mu_0}$$

$\chi = 2.1 \times 10^{-26}$, $B = 5 \times 10^{-5}$ tesla

$$\therefore \quad M = \frac{2.1 \times 10^{-26} \times 5 \times 10^{-5}}{4 \times 10^{-7}}$$

$$= 8.35 \times 10^{-5} \, A/m$$

According to Avogadro's hypothesis, the number of atoms in 32 gm of O_2 is 6.023×10^{23}

Therefore, number of atoms in 1 gm of oxygen is $N = \dfrac{6.023 \times 10^{23}}{32}$

The magnetic dipole moment of each atom is given by

$$\mu = \frac{M}{N} = \frac{8.35 \times 10^{-5} \times 32}{6.023 \times 10^{23}}$$

$$= 44.36 \times 10^{-28} \, A\text{-}m^2 \qquad \text{... Ans.}$$

Example 4.3 : *Calculate magnetic susceptibility of a material assuming one electron and taking $m = 9.1 \times 10^{-31}$ kg, $R = 0.1$ nm , $N = 5 \times 10^{28}/m^3$ and $e = 1.6 \times 10^{-19}$ coulomb.* **(Oct. 17)**

Solution : Magnetic susceptibility of diamagnetic substance is given by

$$\chi = - \frac{\mu_0 N Z e^2}{6m} < R^2 >$$

$$= \frac{- 5 \times 10^{28} \times 4\pi \times 10^{-7} \times (1.6 \times 10^{-19})^2 \times (0.1 \times 10^{-9})}{6 \times 9.1 \times 10^{-31}}$$

$$= - 3 \times 10^{-32} \qquad \text{... Ans.}$$

Example 4.4 : *Nitric oxide is a paramagnetic compound. The magnetic moment of each NO molecule has a maximum component in any direction of about one Bohr magnetron. Compare interaction energy of such magnetic moments in 1.5 T magnetic field with average translational energy of molecule at 300 K.*

Solution : The energy of a magnetic moment in a magnetic field is given by

$$U = -\vec{\mu} \cdot \vec{B} \quad \text{for interaction of } \vec{\mu} \text{ and } \vec{B}$$

Here maximum value of μ is μ_B

$$\text{Magnitude of } U = \mu_B B$$
$$= (9.27 \times 10^{-24} \text{ J/T}) \times (1.5 \text{ T})$$
$$= 1.4 \times 10^{-23} \text{ J}$$
$$= \mathbf{8.7 \times 10^{-5} \text{ eV}} \qquad \textbf{... Ans.}$$

The average translational energy is given by

$$K = \frac{3}{2} kT$$
$$= \frac{3}{2} (1.38 \times 10^{-23} \text{ J/K})(300 \text{ K})$$
$$= 6.2 \times 10^{-21} \text{ J}$$
$$= \mathbf{0.039 \text{ eV}} \qquad \textbf{... Ans.}$$

At 300 K the magnetic interaction energy is only about 0.2% of the thermal kinetic energy

Summary

1. The magnitude of the magnetization vector is defined as the net magnetic dipole moment per unit volume. $\vec{B} = \mu_0 \vec{M} + \mu_0 \vec{H}$ where μ_0 is permeability of free space and $\vec{H}$ is called magnetic field strength. Unit of B is Wb/m^2 and that of H and M is ampere per meter.

2. Magnetization vector $\vec{M}$ is proportional to $\vec{H}$. For these substances, we can write

$$\vec{M} = \chi \vec{H}$$

where the dimensionless quantity χ is called magnetic susceptibility.

3. Diamagnetic substances are weakly repelled by in the external magnetic field and have negative magnetic susceptibility.

4. Langevin's formula for the susceptibility of the diamagnetic material

$$\chi = - \frac{\mu_0 N Z e^2}{6m} < R^2 >$$

5. Some materials when cooled below certain temperature, called critical temperature (T_c) have zero resistance, such material is said to be superconductor.

6. First characteristic property of superconductor is its electrical resistance. For all practical purposes it is zero below a well defined temperature T_c called as critical or transition temperature.

7. The Meissner effect is the expulsion of a magnetic field from a superconductor during its transition to the superconducting state.

8. At a fixed temperature below the critical temperature, superconducting materials cease to super conduct when all external magnetic field is applied which is greater than the critical magnetic field.

9. In a paramagnetic substance each atom contains permanent magnetic dipole moment. Due to thermal agitation, these dipole moments are randomly oriented in the material giving net magnetic moment zero in the absence of external field.

10. Langevin result for paramagnetic susceptibility

$$\chi = \frac{N\mu^2\mu_0}{kT}$$

11. The ferromagnetism in Fe, Co, Ni and Gd is mainly due to self alignment of groups or atoms carrying a permanent magnetic moments in the same direction.

12. Ferromagnetism occurs when paramagnetic ions in a solid, 'lock' together in a small region in which all magnetic moments are along the same direction. Such a region is called domain.

13. In antiferromgnetic the materials like MnO, the two sub lattices have identical magnetic moments but opposite in direction.

14. In ferromagnetic materials in which the two sub-lattices are not identical, there will be a net effect similar to ferromagnetism but considerably weaker.

15. The Neel temperature, T_N, is the temperature above which an antiferromagnetic material becomes paramagnetic.

16. Ferrites are a class of ferrimagnetic ceramic chemical compounds consisting of mixtures of various metal oxides, usually including iron oxides.

17. The ferrites which have a high coercivity and high remanence after magnetization are called hard ferrites.

Exercise

(A) Short Answer Type Questions :

1. What is magnetization vector $\vec{M}$?

2. State Langevin formula for susceptibility of diamagnetic material.

3. What are domains ?

4. How antiferromagnetic material become paramagnetic ?

5. What is Curie constant ?
6. Define Curie temperature.
7. What is superconducting state ?
8. What is critical temperature of superconductor ?
9. What is Meissner effect ?
10. What is critical magnetic field of superconductor ?
11. What is Neel temperature ?
12. What are ferrites?

(B) Long Answer Type Questions :

1. Write a note on diamagnetic material.
2. Explain Langevin theory of diamagnetic material and show that susceptibility is negative.
3. Obtain Langevin formula for paramagnetic susceptibility.
4. What are superconductors? Mention the important property changes that occur in superconductors. Give some examples of practical uses.
5. Describe Meissner effect.
6. Describe Type-I and Type-II superconductors.
7. Describe critical magnetic field of superconductor.
8. Write a note on ferromagnetic materials. Give their examples.
9. Write a note on antiferromagnetic materials.
10. Write a note on ferrimagnetic materials.
11. Explain why it is desirable to use hard ferromagnetic materials to make permanent magnets.
12. What is hysteresis? What information is obtained from hysteresis loop?
13. What are ferrites ? Give their applications.

(C) Unsolved Problems :

1. The magnetic moment of an electron in the ground state of the hydrogen atom is 1 Bohr magneton. Calculate the induced magnetic moment in a field of 1 Wb/m^2.

   (**Ans.** 1.6×10^{-29} A-m^2)

2. In Bohr's model of hydrogen atom, the electron is in a circular orbit of radius 5.3×10^{-11} m and its speed is 2.2×10^6 m/s. What is the magnitude of the magnetic moment due to the electrons motion ?   (**Ans.** 9.3×10^{-24} A-m^2)

3. The magnetic susceptibility of copper is -0.5×10^{-5}. Calculate the magnetic moment per unit volume in copper when subjected to a field of 10^4 amp/metre.

   (**Ans.** 0.05 A/m)

4. The magnetic field strength in silicon is 1000 ampere/metre. If the magnetic field susceptibility is -0.3×10^{-5}. Calculate the magnetization and flux density of silicon.

   (**Ans.** -3×10^{-3} A/m, 1.257×10^{-7} tesla)

October 2015

1. **Attempt all of the following (one mark each) :** (10)
   (a) Define Miller indices.

   Ans. Refer to Section 1.8 on page 1.17.

   (b) State different characterization techniques.

   Ans. Refer to Section (B) Introduction of characterization techniques on page 2.15.

   (c) What is Fermi energy level ?

   Ans. Refer to Section 3.7 on page 3.18.

   (d) Define magnetization.

   Ans. Refer to Section 4.1 on page 4.2.

   (e) What is Hall effect ?

   Ans. Refer to Section 3.8 on page 3.19.

   (f) What is superconductor ?

   Ans. Refer to Section 4.4 on page 4.7.

   (g) What is co-ordination number ?

   Ans. Refer to Section 1.10.3 on page 1.26.

   (h) What are Miller indices of a plane having the intercepts $(1, \infty, \infty)$ on three X, Y and Z axes ?

   Ans. Refer to Section 1.8 on page 1.19.

   (i) Give two assumptions for classical free electron model.

   Ans. Refer to Section 3.2 on page 3.2.

   (j) In Bragg's diffraction condition if $d = 1.5$ A°, then what is the upper limit of λ for obtaining the first order reflection ?

   Ans. Refer to Example 2.5 on page 2.31.

2. **Attempt any two (five marks each) :** (10)
   (a) Show that the volume of unit cell of the reciprocal lattice is inversely proportional to the volume of unit cell of direct lattice.

   Ans. Refer to Section 1.14 (2) on page 1.38.

   (b) Using Ewald's construction show that Bragg's diffraction condition in reciprocal lattice is exactly equivalent to the condition in direct lattice.

   Ans. Refer to Section 2.5 on page 2.5.

   (c) State and explain Meissner effect.

   Ans. Refer to Section 4.6 on page 4.10.

3. **Attempt any two (five marks each) :** (10)
   (a) In a unit cell of simple cubic structure, find the angle between the normals to pair of planes whose Miller indices are
   (i) (100) and (010) and
   (ii) (121) and (111).

   Ans. Refer to Example 1.10 on page 1.49.

(P.1)

 (b) A BCC crystal is used to measure the wavelength of some X-rays. The Bragg angle for the first order reflection from (110) planes is 20.2°. What is the wavelength ? The lattice parameter of the crystal is 3.15 A°.

Ans. Refer to Example 2.4 on page 2.30.

 (c) Evaluate the temperature at which there is one percent probability that a state with energy 0.4 eV above the Fermi energy will be occupied by an electron.

  (Given : $K_B = 1.38 \times 10^{-23}$ joules/Kelvin).

Ans. Refer to Example 3.4 on page 3.32.

4. **(a)** **Attempt any one (eight marks) :**   **(8)**

 (i) Distinguish between metals, semiconductors and insulators on the basis of band theory of solids.

Ans. Refer to Section 3.12 on page 3.28.

 (ii) What do you mean by ferrimagnetism ? What are soft and hard ferrites ? And state any four applications of ferrites.

Ans. Refer to Section 4.13 on page 4.21. Also refer to Section 4.15 on page 4.22.

(b) **Attempt any one (two marks) :**   **(2)**

 (i) Find number of atoms per unit cell for body centered cubic system.

Ans. Refer to Section 1.10.2 (3) on page 1.25.

 (ii) A monochromatic beam of X-rays having wavelength 1.2 A° is incident on a crystal. The first order maxima of reflected X-rays are obtained at an angle of 7°. Calculate the glancing angle for second order reflection.

Ans. Refer to Example 2.1 on page 2.28.

❑❑❑

April 2016

1. **Attempt all of the following (one mark each) :**   **(10)**

 (a) Define packing fraction.

Ans. Refer to Section 1.11 on page 1.26.

 (b) Why conductivity of semiconductors increases with temperature ?

Ans. Refer to Section 3.2 on page 3.2.

 (c) Give any two applications of superconductivity.

Ans. Refer to Section 4.6 on page 4.13.

 (d) Define crystal lattice.

Ans. Refer to Section 1.2 on page 1.3.

 (e) Give Bragg's diffraction condition for direct lattice.

Ans. Refer to Section 2.6 on page 2.7.

 (f) What is mobility ?

 (g) State Bloch theorem.

Ans. Refer to Section 3.5 on page 3.14.

(h)   Determine the number of atoms per unit cell for face centered cubic cell.

Ans. Refer to Section 1.10.2 (2) on page 1.25.

(i)   What do you mean by domains in ferromagnetic materials ?

Ans. Refer to Section 4.10 on page 4.18.

(j)   Give any two uses of Thermal Gravimetric Analysis (TGA).

Ans. Refer to Section 2.9 on page 2.17.

2. Attempt any two (five marks each) :   **(10)**

(a)   Obtain an expression for interplaner spacing for simple cubic system.

Ans. Refer to Section 1.9, Case-1 on page 1.21.

(b)   Write a short note on ultra-violet and visible absorption spectroscopy.

Ans. Refer to Section 2.10 on page 2.19.

(c)   What are ferrites ? Give any two examples and six applications of ferrites.

Ans. Refer to Section 4.15 on page 4.22.

3. Attempt any two (five marks each) :   **(10)**

(a)   Calculate the Miller indices of crystal planes, which cut through the crystal axes at
       (i) (6a, 3b, 3c) and (ii) (2a, −3b, −3c).

Ans. Refer to Example 1.8 on page 1.46.

(b)   The distance between (111) planes in a face centered cubic crystal is 2A °.
       Determine the lattice parameter and atomic diameter.

Ans. Refer to Example 2.3 on page 2.30.

(c)   In a Hall effect experiment on zinc, a potential of 4.5 µV is developed across a foil
       of thickness 0.02 mm when a current of 1.5 A is passed in a direction perpendicular
       to a magnetic field 2.0 T.
       Calculate (i) Hall coefficient for zinc and (ii) Electron density.

Ans. Refer to Example 3.8 on page 3.35.

4. (A) Attempt any one (eight marks) :   **(8)**

(i)   State three assumptions of Sommerfeld's free electron model and obtain an
       expression for energy levels and density of states in one dimension.

Ans. Refer to Section 3.3 on page 3.6.

(ii)   What is paramagnetism ? Obtain Langevin's formula for paramagnetic
       susceptibility.

Ans. Refer to Sections 4.7 and 4.8 on page 4.13 and 4.15.

(B) Attempt any one (two marks) :   **(2)**

(i)   Sketch (111) and (101) planes in simple cubic cell.

Ans. Refer to Example 1.4 and 1.5 on page 1.43 and 1.44.

(ii)   Calculate the distance between two lattice planes, which give first order diffraction
       at an angle of 26.42° with X-rays of wavelength 0.75 A°.

Ans. Refer to Example 2.8 on page 2.32.

❏❏❏

October 2016

1. Attempt all of the following (One mark each) :   **(10)**

(a)   Define the term 'Space lattice'.

Ans. Refer to Section 1.3 on page 1.4.

(b)   What are domains ?

Ans. Refer to Section 4.10 on page 4.18.

(c)   What is mean free path ?

Ans. Refer to Section 3.2 on page 3.3.

(d)   State any two uses of thermal gravimetric analysis technique.

Ans. Refer to Section 2.9 on page 2.17

(e)   Define the term 'Packing fraction'.

Ans. Refer to Section 1.11 on page 1.26.

(f)   State Bragg's diffraction condition in direct lattice.

Ans. Refer to Section 2.4 on page 2.4.

(g)   State Hall effect.

Ans. Refer to Section 3.8 on page 3.19.

(h)   State any two spectroscopic techniques for the analysis of crystal structure.

Ans. Refer to Section 2.8 (B) on page 2.15.

(i)   Give formula for Fermi function.

Ans. Refer to Section 3.7 on page 3.18.

(j)   Why ordinary optical grating cannot diffract X-rays ?

Ans. Refer to Section 2.3 on page 2.3.

2. Attempt any two (Five marks each) :   **(10)**

(a)   Show that every reciprocal lattice vector is perpendicular to a direct lattice plane.

Ans. Refer to Section 1.14 (1) on page 1.37.

(b)   Write short note on scanning tunnelling electron microscopy.

Ans. Refer to Section 2.12.1 on page 2.27.

(c)   Write a short note on domain and hysteresis of ferromagnetic materials.

Ans. Refer to Section 4.10 on page 4.18. Also refer to Section 4.11 on page 4.19.

3. Attempt any two (Five marks each) :   **(10)**

(a)   Find out the number of atoms per square millimeter on a plane (100) of lead whose interatomic distance is 3.449 A°. Lead has FCC structure.

Ans. Refer to Example 1.2 on page 1.42.

(b)   In case of certain crystal the maxima of reflected X-rays are obtained at glancing angles 5°, 12', 7° 24' and 9° 1' respectively from three different reflecting planes. Determine the types of cubic crystal.

Ans. Refer to Example 2.2 on page 2.29.

(c)   The Fermi energy of copper is 7.1 eV. Assuming that it is the maximum kinetic energy of electrons in copper. Find the number of atoms per unit volume in copper. (Given : Mass of electron $(m) = 9.1 \times 10^{-31}$ kg,

Planck's constant $(h) = -6.62 \times 10^{-34}$ Joule-sec).

Ans. Refer to Example 3.1 on page 3.31.

4. (a) Attempt any one (Eight marks) : (8)
    (i)   Obtain an expression for energy levels and density of states in one dimension.
    Ans. Refer to Section 3.4 on page 3.6.
    (ii)   What is diamagnetism ? Obtain classical Langevin's formula for the susceptibility of the diamagnetic material.
    Ans. Refer to Sections 4.2 and 4.3 on page 4.3 and 4.4.
    (b) Attempt any one (Two marks) : (2)
    (i)   Find the number of atoms per unit cell for BCC crystal structure.
    Ans. Refer to Section 1.10.2 (3) on page 1.25.
    (ii)   Draw the schematic labelled diagram of UV-visible spectrophotometer.
    Ans. Refer to Section 2.11.2 on page 2.22.

❑❑❑

April 2017

1. Attempt all of the following (One mark each) : (10)
    (a)   Define the term 'symmetry operation'.
    Ans. Refer to Section 1.6 on page 1.8.
    (b)   What is mobility ?
    Ans. Refer to Summary (point 14) on page 3.37.
    (c)   State Bragg's diffraction condition in reciprocal lattice.
    Ans. Refer to Section 2.5 on page 2.5.
    (d)   State any two assumptions of classical free electron model.
    Ans. Refer to Section 3.2 on page 3.2.
    (e)   Define the term 'Fermi energy'.
    Ans. Refer to Section 3.7 on page 3.18.
    (f)   What is Neel temperature ?
    Ans. Refer to Section 4.14 on page 4.21.
    (g)   Define the term 'spectroscopy'.
    Ans. Refer to Section 2.10 on page 2.19.
    (h)   What is meant by fold number ?
    Ans. Refer to Section 1.10.3 on page 1.26.
    (i)   If the angle between the diffraction of incident X-rays and diffracted one is 20°, what is the angle of incidence ?
    Ans. Refer to Example 2.1 on page 2.28.
    (j)   State any two applications of scanning electron microscopy.
    Ans. Refer to Section 2.12 on page 2.25.

2. Attempt any two (Five marks each) : (10)
    (a)   Describe the Zinc blende (ZnS) and hexagonal closed packed (hcp) structures with the help of neat diagrams.
    Ans. Refer to Section 1.12 (4) and (5) on page 1.30 and 1.31.
    (b)   Write short note on Thermal Gravimetric Analysis (TGA).
    Ans. Refer to Section 2.9 on page 2.17.

   (c)   State and explain Meissner effect.

Ans. Refer to Section 4.6 on page 4.10.

3. Attempt any Two (Five marks each) :   **(10)**

   (a)   Show that FCC lattice is the reciprocal of the BCC lattice.

Ans. Refer to Sections 1.10.1 and 1.10.2 on page 1.23 and 1.25.

   (b)   An X-ray analysis of crystal is made with monochromatic X-rays of wavelength 0.6 A°. Bragg's reflections are obtained at angles of :

        (i) 6.45°, (ii) 9.15° and (iii) 13°.

        Calculate the interplaner spacing of the crystal.

Ans. Refer to Example 2.7 on page 2.32.

   (c)   Evaluate the temperature at which there is 1% probability that a state with an energy 0.4 eV above the Fermi energy will be occupied by an electron.

        (Given : $K_B = 1.38 \times 10^{-23}$ Joule/Kelvin).

Ans. Refer to Example 3.4 on page 3.32.

4. (a)   Attempt any One (Eight marks) :   **(8)**

   (i)   Obtain expressions for density of states and Fermi energy at absolute zero in three dimensions.

Ans. Refer to Section 3.4.3 on page 3.11. Also refer to Section 3.7 on page 3.18.

   (ii)   What is paramagnetism ? Obtain Langevin's formula for the susceptibility of the paramagnetic material.

Ans. Refer to Sections 4.7 and 4.8 on page 4.13 and 4.15.

(b)   Attempt any One (Two marks) :   **(2)**

   (i)   Find the value of packing fraction for simple cubic system.

Ans. Refer to Section 1.11 (1) on page 1.27.

   (ii)   Draw the schematic labelled diagram of scanning electron microscope.

Ans. Refer to Section 2.12 on page 2.25.

❑❑❑

October 2017

1. Attempt all of the following (One mark each) :   **(10)**

   (a)   Define packing fraction.

Ans. Refer to Section 1.11 on page 1.26.

   (b)   State Bragg's diffraction condition in direct lattice.

Ans. Refer to Section 2.4 on page 2.4.

   (c)   Give any two applications of SEM.

Ans. Refer to Section 2.12 on page 2.25.

   (d)   Give the names of two characterization techniques.

Ans. Refer to Section Introduction on page 2.15.

   (e)   Define mean free path.

Ans. Refer to Section 3.2 on page 3.3.

   (f)   State Hall effect.

Ans. Refer to Section 3.8 on page 3.19.

(g)   Define mobility.

(h)   Define Miller indices.

Ans. Refer to Section 1.8 on page 1.17.

(i)   What is Neel's temperature ?

Ans. Refer to Section 4.14 on page 4.21.

(j)   Define superconductivity.

Ans. Refer to Section 4.5 on page 4.8.

2.   **Attempt any two (Five marks each) :**   **(10)**

(a)   Obtain an expression for interplanar distance. Hence show that for simple cubic

system, $d_{hkl} = \dfrac{a}{\sqrt{h^2 + k^2 + l^2}}$.

Ans. Refer to Section 1.9 on page 1.19.

(b)   Show that Bragg's diffraction condition in the reciprocal lattice is $2\vec{K} \cdot \vec{G} + G^2 = 0$.

Ans. Refer to Section 2.5 on page 2.5.

(c)   Write a note on type-I and type-II superconductors.

Ans. Refer to Section 4.6 on page 4.10.

3.   **Attempt any two (Five marks each) :**   **(10)**

(a)   In a unit cell of simple cubic structure, find the angle between the normals to the
pair of planes whose Miller indices are

(i)   (100) and (010) and

(ii)   (121) and (111).

Ans. Refer to Example 1.10 on page 1.49.

(b)   X-rays with $\lambda = 1.54$ A° are used to calculate the spacing of (200) planes in
aluminium. The Bragg's angle for first order reflection is 22.4 A°. What is the size of
unit cell of the aluminium crystal ?

Ans. Refer to Example 2.5 on page 2.31.

(c)   Calculate magnetic susceptibility of a material assuming one electron and taking
$m = 9.1 \times 10^{-31}$ kg, $R = 0.1$ mm, $N = 5 \times 10^{28}/m^3$, and $e = 1.6 \times 10^{-19}$ coulomb.

Ans. Refer to Example 4.3 on page 4.25.

4.   (a)   **Attempt any one (Eight marks each) :**   **(8)**

(i)   Write detailed note on TGA and SEM.

Ans. Refer to Section 2.9 on page 2.17 and Section 2.12 on page 2.25.

(ii)   Obtain an expression for energy levels and density of states in one dimension.

Ans. Refer to Section 3.4 on page 3.6.

(b)   **Attempt any one (Two marks each) :**   **(2)**

(i)   Sketch (100) and (101) planes in simple cubic cell.

Ans. Refer to Section 1.8 on page 1.19.

(ii)   What are ferrites ? Give two examples.

Ans. Refer to Section 4.15 on page 1.23.

❏❏❏

April 2018

1. Attempt all of the following (One mark each) :   **(10)**
- (a) Define packing fraction.
- (b) Define the term symmetry operation.
- (c) State Bragg's diffraction condition in reciprocal lattice.
- (d) What is nearly free electron model ?
- (e) What do you mean by quantitative analysis ?
- (f) Define Fermi energy.
- (g) What are domains ?
- (h) What is Neel temperature ?
- (i) Define super conductors.
- (j) Define primitive unit cell.

2. Attempt any two of the following (5 marks each) :   **(10)**
- (a) Discuss crystal structure of NaCl in detail.
- (b) With the help of Ewald's construction, show that the diffraction condition in reciprocal lattice is exactly equivalent to 2d sin θ = nλ in the direct lattice.
- (c) Write a note on type-I and type-II superconductors.

3. Attempt any two of the following (5 marks each) :   **(10)**
- (a) Find out the number of atoms per square millimeter on a plane (100) of lead whose interatomic distance is 3.499 A°. Lead has face-centred cubic structure.
- (b) In a Hall effect experiment on Zinc, a potential of 4.5 μV is developed across a foil of thickness 0.02 mm. When a current of 1.5 A is passed in a direction perpendicular to a magnetic field of 2.0 T, calculate :
  - (i) The Hall coefficient for zinc.
  - (ii) The electron density. (Given : Charge on electron = 1.6×10^{-19} C).
- (c) A paramagnetic substance has 10^{28} atoms/m^3. The magnetic moment of each atom is 1.79×10^{-23} A-m^2. Calculate the paramagnetic susceptibility of the material at temperature 320°K. What would be the dipole moment of the rod of this material 0.1 m long and 1 cm^2 cross-section placed in a field of 7×10^4 A/m ? (K = 1.38×10^{23} J/°K, $\mu_0 = 4\pi \times 10^{-7}$ Wb/A-m).

4. (a) Attempt any one of the following (eight marks) :   **(8)**
- (i) State three assumptions of Sommerfeld's free electron model and obtain an expression for energy levels and density of states in one dimension.
- (ii) Write a detailed note on TGA and ultraviolet visible spectrophotometer.

(b) Attempt any one of the following (two marks) :   **(2)**
- (i) What are ferrites ? Give two examples.
- (ii) Sketch (112) and (2, 0, 0) planes in simple cubic cell.

□□□